THE NATIONAL ACADEMIES

Advisers to the Nation on Science, Engineering, and Medicine

The **National Academy of Sciences** is a private, nonprofit, self-perpetuating society of distinguished scholars engaged in scientific and engineering research, dedicated to the furtherance of science and technology and to their use for the general welfare. Upon the authority of the charter granted to it by the Congress in 1863, the Academy has a mandate that requires it to advise the federal government on scientific and technical matters. Dr. Ralph J. Cicerone is president of the National Academy of Sciences.

The **National Academy of Engineering** was established in 1964, under the charter of the National Academy of Sciences, as a parallel organization of outstanding engineers. It is autonomous in its administration and in the selection of its members, sharing with the National Academy of Sciences the responsibility for advising the federal government. The National Academy of Engineering also sponsors engineering programs aimed at meeting national needs, encourages education and research, and recognizes the superior achievements of engineers. Dr. Charles M. Vest is president of the National Academy of Engineering.

The **Institute of Medicine** was established in 1970 by the National Academy of Sciences to secure the services of eminent members of appropriate professions in the examination of policy matters pertaining to the health of the public. The Institute acts under the responsibility given to the National Academy of Sciences by its congressional charter to be an adviser to the federal government and, upon its own initiative, to identify issues of medical care, research, and education. Dr. Harvey V. Fineberg is president of the Institute of Medicine.

The **National Research Council** was organized by the National Academy of Sciences in 1916 to associate the broad community of science and technology with the Academy's purposes of furthering knowledge and advising the federal government. Functioning in accordance with general policies determined by the Academy, the Council has become the principal operating agency of both the National Academy of Sciences and the National Academy of Engineering in providing services to the government, the public, and the scientific and engineering communities. The Council is administered jointly by both Academies and the Institute of Medicine. Dr. Ralph J. Cicerone and Dr. Charles M. Vest are chair and vice chair, respectively, of the National Research Council.

www.national-academies.org

NATIONAL RESEARCH COUNCIL COMMITTEE ON COUNTERTERRORISM CHALLENGES FOR RUSSIA AND THE UNITED STATES

Siegfried S. Hecker, Director Emeritus, Los Alamos National Laboratory; Co-Director and Professor, Center for International Security and Cooperation, Stanford University, *Chair*
Wm. A. Wulf, President, National Academy of Engineering, *Ex-officio*
Robert McC. Adams, Adjunct Professor, University of California at San Diego
John F. Ahearne, Director, Ethics Program, Sigma Xi, The Scientific Research Society
Lewis M. Branscomb, Aetna Professor of Public Policy and Corporate Management, Emeritus, John F. Kennedy School of Government, Harvard University
George Bugliarello, President Emeritus and University Professor, Polytechnic University
Anita K. Jones, Lawrence R. Quarles Professor of Engineering and Applied Science, University of Virginia
Michael Moodie, Independent Consultant and former President, Chemical and Biological Arms Control Institute
Russ Zajtchuk, President, Chicago Hospitals International

Staff

Glenn E. Schweitzer, Program Director, National Research Council
Kelly Robbins, Senior Program Officer, National Research Council
A. Chelsea Sharber, Senior Program Associate, National Research Council

COUNTERING TERRORISM

Biological Agents, Transportation Networks, and Energy Systems

Summary of a U.S.–Russian Workshop

Glenn E. Schweitzer, *Rapporteur*

Committee on Counterterrorism Challenges for
Russia and the United States

Office for Central Europe and Eurasia
Development, Security, and Cooperation
Policy and Global Affairs

NATIONAL ACADEMY OF SCIENCES
THE NATIONAL ACADEMIES

In cooperation with the Russian Academy of Sciences

THE NATIONAL ACADEMIES PRESS
Washington, D.C.
www.nap.edu

THE NATIONAL ACADEMIES PRESS 500 Fifth Street, N.W. Washington, DC 20001

NOTICE: The project that is the subject of this report was approved by the Governing Board of the National Research Council, whose members are drawn from the councils of the National Academy of Sciences, the National Academy of Engineering, and the Institute of Medicine. The members of the committee responsible for the report were chosen for their special competences and with regard for appropriate balance.

This study was supported by Grant No. B 7075.R03 between the National Academy of Sciences and the Carnegie Corporation of New York. Any opinions, findings, conclusions, or recommendations expressed in this publication are those of the author(s) and do not necessarily reflect the views of the organizations or agencies that provided support for the project.

International Standard Book Number-13: 978-0-309-12707-3
International Standard Book Number-10: 0-309-12707-6

A limited number of copies are available from the Office for Central Europe and Eurasia, National Research Council, 500 Fifth Street, N.W., Washington, DC 20001; (202) 334-2376.

Additional copies of this report are available from the National Academies Press, 500 Fifth Street, N.W., Lockbox 285, Washington, DC 20055; (800) 624-6242 or (202) 334-3313 (in the Washington metropolitan area); Internet, http://www.nap.edu.

Printed in the United States of America.

RUSSIAN ACADEMY OF SCIENCES STANDING COMMITTEE ON COUNTERTERRORISM

Academician Yevgeny Velikhov, Director, Russian Research Center—Kurchatov Institute, *Chair*

RAS Corresponding Member Leonid Bolshov, Director, Russian Academy of Sciences Nuclear Safety Institute

Academician Nikolay Laverov, Vice President, Russian Academy of Sciences

Academician Nikolay Platé, Vice President, Russian Academy of Sciences *(deceased)*

Academician Aleksander Spirin, Director, Russian Academy of Sciences Protein Institute

Academician Konstantin V. Frolov, Director, Russian Academy of Sciences Institute of Machine Science *(deceased)*

RAS Corresponding Member Valery Tishkov, Director, Russian Academy of Sciences Institute of Ethnology and Anthropology

Mr. Gennady Kovalenko, Presidium of the Russian Academy of Sciences

Dr. Renat S. Akchurin, Chief of the Cardiovascular Surgery Department, Cardiology Research Center

Staff

Yury K. Shiyan, Chief Expert, Head of the Desk on Cooperation with North and Latin American Countries, Foreign Relations Department

Preface

This report presents the proceedings of the fourth U.S.-Russian interacademy workshop on the general theme of countering terrorism, which was held in Moscow in March 2007. The first report was published by the National Academy Press (now the National Academies Press) in 2002 under the title *High Impact Terrorism: Proceedings of a Russian-American Workshop*. The second report was published in 2004 under the title *Terrorism: Reducing Vulnerabilities and Improving Responses: U.S.-Russian Workshop Proceedings*. The third report was published in 2006 under the title *Countering Urban Terrorism in Russia and the United States: Proceedings of a Workshop*. The present report continues to explore topics related to urban terrorism but with a new emphasis on potential attacks involving biological agents, transportation networks, and energy systems. The Carnegie Corporation of New York has generously supported all four of the workshops and the preparation of the reports.

Two other recent projects carried out as cooperative efforts of the National Academies and the Russian Academy of Sciences also deserve mention. They were closely linked to the activities reflected in the reports of the four workshops noted above. These two projects resulted in consensus reports also published by the National Academies Press entitled *Biological Science and Biotechnology in Russia: Controlling Diseases and Enhancing Security*, 2006, and *U.S.-Russian Collaboration in Combating Radiological Terrorism*, 2007.

This report is organized into several sections. First, summary reports of discussions at meetings of three interacademy working groups that were convened just before the workshop are presented. These working groups addressed bioterrorism, transportation vulnerabilities, and energy system vulnerabilities. The agendas for the working groups and plenary sessions are included in Appendix A, along with a list of participants. Second, 18 papers that provided the basis for presentations during the working group discussions and workshop plenary sessions are set forth in their entirety. Appendix B identifies some important recent books and reports published in Russia that are highly relevant to the topics that were discussed. Finally, the presentation by a senior Russian government official, included in Appendix C, provides important perspectives that were taken into account during the discussions. We hope that the discussions at the workshop will be of assistance to the two governments as they continue to develop the governmental frameworks for combating urban terrorism and approaches to international collaboration in this field.

The amount of information that was exchanged during the meetings of the working groups, the plenary session of the workshop, and the related visits to organizations and facilities in the Moscow and St. Petersburg areas was extensive. The sampling of information presented in this report underscores the value of international exchanges in the rapidly expanding field of counterterrorism. Much of the information is of direct relevance in efforts to enhance national security awareness in both countries, and indeed throughout the international community.

ACKNOWLEDGMENTS

This publication was made possible by a grant from the Carnegie Corporation of New York. The Russian Academy of Sciences, in cooperation with other Russian organizations, did an excellent job in arranging all aspects of the visit to Russia. We express our sincere appreciation to the Carnegie Corporation and to all of the Russian organizations that were involved for their assistance.

The statements made and views expressed in this report are solely the responsibility of the authors and the rapporteur. They do not necessarily represent the positions of the planning committees, the Carnegie Corporation, the National Academies, the Russian Academy of Sciences, or other organizations where the authors are employed.

This volume has been reviewed in draft form by individuals chosen for their technical expertise, in accordance with procedures approved by the National Research Council's Report Review Committee. The purpose of this independent review is to provide candid and critical comments that will assist the institution in making its published report as sound as possible and to ensure that the report meets institutional standards for quality. The review comments and draft manuscript remain confidential to protect the integrity of the process.

We wish to thank the following individuals for their review of selected papers and summary material: Edward Badolato, Integrated Infrastructure Analytics, Inc.; Kavita Berger, American Association for the Advancement of Science; Mortimer Downey, PB Consult, Inc; Robert Gallamore, Northwestern University; Sanjay Jain, The George Washington University; Brian Lopez, Lawrence Livermore National Laboratory; Neil Smelser, University of California, Berkeley; Amy Smithson, Monterey Institute of International Studies; Theofanis Theofanous, University of California, Santa Barbara; Alvin Trivelpiece, Retired, Oak Ridge National Laboratory; and Wm. A. Wulf, University of Virginia.

Although the reviewers listed above have provided constructive comments and suggestions, they were not asked to endorse the content of the individual papers. Responsibility for the final content of the volume rests with the individual authors and the rapporteur.

Special appreciation is extended to Kelly Robbins for her translation of many of the Russian language papers into English and for her assistance in editing this report. Also, we appreciate the work of Jan Dee Summers in editing this volume.

Siegfried S. Hecker
Chair, Committee on Counterterrorism Challenges for Russia and the United States of the National Research Council

Glenn E. Schweitzer
Rapporteur
Director, Office for Central Europe and Eurasia
National Research Council

Contents

Summaries

1 U.S.-Russian Working Group on Bioterrorism 3
Claire Cornelius (Rapporteur)

2 U.S.-Russian Working Group on Transportation System Vulnerabilities 7
Cynthia Getner (Rapporteur)

3 U.S.-Russian Working Group on Energy System Vulnerabilities 14
A. Chelsea Sharber (Rapporteur)

Selected Papers

Overview

4 Tendencies in Global Terrorism 25
Raphael F. Perl

5 Use of Predictive Modeling Packages for Effective Emergency Management 32
Nikolai P. Kopylov and Irek R. Khasanov

6 Organizational Measures and Decision Support Systems for Preventing and Responding to Terrorist Acts at Potentially Hazardous Facilities, on Transportation Systems, and in Locations Where Large Numbers of People Congregate 46
A. Yu. Kudrin, A. I. Zaporozhets, and S. A. Kachanov

7 Characteristics of Technological Terrorism Scenarios and Impact Factors 53
Nikolai A. Makhutov, Vitaly P. Petrov, and Dmitry O. Reznikov

8 Activities of the Russian Federal Medical-Biological Agency Related to Radiation, Chemical, and Biological Security 70
Vladimir V. Romanov

Bioterrorism

9 Disease Surveillance and International Biosecurity 73
David R. Franz

10 Emerging Viral Infections in the Asian Part of Russia 79
Sergei V. Netesov and Natalya A. Markovich

Transportation Vulnerabilities

11 A Note on the Interfacial Vulnerabilities of Transportation Systems 95
George Bugliarello

12 Transportation Planning for Evacuations 104
John C. Falcocchio

13 International and National Priorities in Combating Terrorism in the Transportation Sector 116
Vladimir N. Lopatin

Energy System Vulnerabilities

14 Managing the Radius of Risk 124
Drew F. Lieb

15 The Problem of Oil and Natural Gas Pipeline Security 150
Sergei G. Serebryakov

Other Counterterrorism Topics

16 U.S.-Russian Collaboration in Combating Radiological Terrorism 160
John F. Ahearne

17 IAEA Activities in Preventing Radiological and Nuclear Terrorism 173
Miroslav Gregoric

18 Electromagnetic Terrorism: Threat to the Security of the State Infrastructure 186
Vladimir Ye. Fortov and Yury V. Parfyonov

19 The Phenomenon of Suicide Bombings in Israel: Lessons Learned 189
Mordecai Z. Dzikansky

20 Raman Spectroscopic Detection of Chemical, Biological, and Explosive Agents 200
Russ Zajtchuk and Gary R. Gilbert

21 The U.S. Department of Homeland Security Science and Technology Directorate 208
John F. O'Neil

Appendixes

A Agenda 217

B Recent Russian and International Publications of Interest 224

C Russia's Counterterrorism Strategy 226
Valentin A. Sobolev

Summaries

1

U.S.-Russian Working Group on Bioterrorism

Claire Cornelius (Rapporteur)

Experts from several U.S. and Russian organizations convened March 19-20, 2007, at the Russian Academy of Sciences (RAS) in Moscow to discuss terrorism threats and responses involving biological pathogens and to further define the role that the scientific, medical, and agricultural communities should play in preventing and containing bioterrorist occurrences in the United States, Russia, and worldwide. Informal presentations were made by specialists in the areas of biomedical research, epidemiology, public health, and scientific instrumentation technology as indicated in Appendix A. The American specialists also had the opportunity to visit several Russian health-oriented organizations in Moscow and St. Petersburg as discussed below.

SITE VISITS

Members of the working group met with representatives from the Federal Medical–Biological Agency (FMBA) and the Center for Hygiene and Epidemiology (Rospotrebnadzor) in Moscow.

In 2004, under a new administrative reform, several research institutions were transferred from the Ministry of Health to FMBA, including the important Research Institute of Highly Pure Biopreparations in St. Petersburg. Four major functions of FMBA are (1) organization and implementation of state sanitary and epidemiological services for industries and territories with known dangerous working conditions; (2) detection and containment of hazardous agents (nuclear, biological, and chemical); (3) development of policy and regulations governing the safety and well-being of the population, with emphasis on those employed in industries and institutions with dangerous working conditions; and (4) provision of direct medical care for researchers and industry workers working in dangerous fields, including those who previously were affiliated with activities of Biopreparat. Various programs and technological capabilities were discussed, particularly in the areas of chemical research, which is a major concern of FMBA. Russian specialists at FMBA emphasized the need for more formalized biological and chemical safety programs for workers and appropriate medical services and financial compensation when appropriate for researchers and laboratory workers, while others emphasized the need for general nonproliferation strategies for addressing weapons of mass destruction.

The discussion with the specialists of the Center for Hygiene and Epidemiology focused on (1) methods of surveillance and control of communicable disease, (2) areas of public health and medical research of special interest, and (3) model development for predictions of morbidity and mortality. In response to a bioterrorist incident or significant infectious disease outbreak, the center plays a key role in cooperation with other public health organizations in Moscow. The visitors were able to observe the operation of the computer-based epidemiological program, which is connected to hospitals and other public health facilities throughout the city as well as to the federal ministries and agencies responsible for public health. At present, this center is not a partner in an international epidemiological network for megacities, but it could make significant contributions to such a network.

In St. Petersburg, the Research Institute of Highly Pure Biopreparations has achieved Good Laboratory Practices (GLP) and Good Manufacturing Practices (GMP) status with the assistance of American collaborators with research and production experience who served as consultants funded by the U.S. Department of State. At present, only a very small percentage of the institute's income comes from international contracts and grants (less than 2 percent, in contrast to much higher levels in years past). Sales from the production of several locally developed drugs for consumption in Russia complement the core budget and the income from local contracts and grants. At the same time, the institute is highly motivated to retain contacts with American colleagues in order to stay abreast of worldwide developments in research and related activities.

Also in St. Petersburg, a visit to the Institute of Influenza focused on the response of Russia to the outbreak of avian influenza, particularly in the Siberian region of the country. The institute serves as the hub for the national effort to respond to the outbreak. The response was complicated by the large number of small poultry operations scattered over large geographical areas, which required the killing of tens of thousands of birds—both domestic and wild. Institute representatives emphasized the value of international collaboration, through the World Health Organization and other bilateral and international mechanisms, as essential in efforts to control this outbreak as well as other types of influenza that often cross international borders. (See Chapter 10 for more details on recent avian influenza outbreaks and control strategies in Russia.)

A brief meeting with public health authorities of the city of St. Petersburg provided opportunities to consider the relationships between preparations for responding to a bioterrorism incident and ongoing activities directed to control of public health problems. The leadership for such response is vested within the Office of the Mayor, which can mobilize support from a wide variety of research and other scientific facilities throughout the city. The Security Committee of the city, which had just moved into very modern new premises with excellent communication facilities, has overall responsibility for overseeing all such incidents. Public health and scientific institutes are represented on the committee. To date, the only threat apparently has been a continuing influx of hoax letters—more than 1,000 in 2006—containing harmless white powder that have been received by various government organizations.

AREAS OF COMMON INTEREST

The roundtable discussions in Moscow reviewed key issues surrounding bioterrorism, and particularly developments in countermeasures in both countries, that have arisen since the previous workshop in 2005. At the onset of the workshop, several participants urged the scientific and medical communities not to lose sight of the fact that nature has a prodigious arsenal of bioagents (some known, some not, most zoonotic) that are deleterious in their own right to humankind. Of special concern are multidrug-resistant pathogens and pathogens for which there are no vaccines or therapeutic remedies, as well as nosocomial infections.

Working group members pointed out that the challenge for the scientific community worldwide is multifaceted. Some focused on the need for the scientific community to make every effort to discover causative agents of disease with alacrity and great precision. To this end, highly sensitive detection and sample collection tools and devices should be available and a solid global team approach should be instituted. Within the context of international teamwork, the working group examined the strengths and weaknesses of previous collaboration strategies (the Cooperative Threat Reduction Program of the U.S. Department of Defense, the International Science and Technology Center, the U.S. Civilian Research and Development Foundation, the BioIndustry Initiative of the U.S. Department of State, the Nuclear Threat Initiative, and so forth). Several discussants stated that the focus of research and development on technical applications, preventive measures, and therapeutics must be comprehensive and take into consideration all of the intricacies and niches of the biological agents of disease, as emerging biothreats are often species neutral. Additionally, an efficient and secure epidemiological data collection system should be developed to facilitate information reporting and exchange. Finally and most importantly, they felt that scientists and medical professionals must continue to draw on their skills and expertise to positively influence the way governments shape biodefense and biodisaster policies.

Although commonality in data collection and management is desired, it is not an easy feat to accomplish. In fact, some working group participants identified an even larger concern—that is, the extent to which scientists can safely acquire and disseminate the data. Joint projects and a certain degree of transparency benefit many researchers and communities; however, this same approach poses the potential threat of negative utilization and exploitation. These individuals emphasized the need for standard operating procedures worldwide for the handling of pathogens, along with criteria for the sharing of sensitive (and potentially deadly) scientific discoveries. Additionally, a global consensus is needed on a code of conduct of scientists engaged in biological research, particularly those handling "select agents." The importance of education and specialized training for biomedical researchers and their professional development and recruitment and retention was highlighted by some working group members: We should not forget the value of increasing the scientific literacy of the general populace, which in turn can assist in efforts to recognize and respond appropriately to a biological catastrophe as well as ensure sustained scientific scholarship through public support.

Working group members reaffirmed the value of international collaboration in assisting in the retention of biological scientists in Russia essential to strengthening the scientific workforce, strengthening the biological nonproliferation regime by promoting transparency and fostering trust among scientists, and generating joint awareness and reporting of infectious diseases. Several members also suggested potential future joint initiatives in expanded biosafety and biosecurity activities, development of response technologies, development of medical countermeasures; and joint studies of emerging infections.

Throughout the discussions, the members of the working group noted the importance of linking national activities with broader global programs. In particular, the roles of the World Health Organization, the Food and Agriculture Organization of the United Nations, and the World Organization for Animal Health were acknowledged.

2

U.S.-Russian Working Group on Transportation System Vulnerabilities

Cynthia Getner (Rapporteur)

The National Academies–Russian Academy of Sciences Working Group on Transportation System Vulnerabilities met in Moscow March 19-20, 2007, to exchange information on vulnerabilities as they relate to urban terrorism. The Institute of Machine Sciences of the Russian Academy of Sciences (RAS) hosted the meetings, which were followed by site visits in Moscow to two facilities of the Ministry for Civil Defense, Emergencies, and Elimination of Consequences of Natural Disasters (EMERCOM)—the Research Institute for Civil Defense and Disaster Management and the Research Institute for Fire Protection. Additional visits to EMERCOM facilities were carried out in St. Petersburg as indicated in Appendix A.

WORKING GROUP PRESENTATIONS

Academician Konstantin Frolov began the discussion with a presentation on the scientific basis for countering terrorism aimed at urban transportation. He focused on the human factor, which must be taken into account in both terrorism and other types of disasters. An in-depth analysis carried out by Nikolai Makhutov and his staff at the Institute of Machine Sciences indicated the extent to which the human factor is important in disasters and accidents as well as to vulnerabilities that can be exploited by terrorists.

To put terrorism in perspective, in Russia approximately 200 people lose their lives to terrorism annually, while more than 20,000 lose their lives to fire, 30,000-35,000 deaths are attributable to road accidents, and about 15,000 die from water accidents.

Academician Frolov described a framework for reducing vulnerabilities that begins with the Security Council of the Russian Federation. It then has two organizational paths: (1) a scientific panel of the Security Council and (2) various federal departments. Both paths lead to subordinate agencies and organizations that contribute their expertise and analysis. A joint strategy is developed that results in technical standards and rules that are articulated, certified as technically sound, and then accredited as requirements.

During the discussion, participants noted the importance of the human factor when dealing with Hurricane Katrina and the relevance of this experience in responding to terrorism. Of special interest was the involvement of the security services when dealing with terrorism.

In his presentation on control and supervision as a prerequisite for ensuring the safety of transportation systems, Vladimir Chertok of the Federal Transportation Supervision Service stressed that a unified system of safety assurance for all transportation modes is needed. Standardized protocols and practices involving different modes of transportation are particularly important. At this time, the Federal Authority for Transport Oversight supervises all types of transport. In Russia, 120 subnational entities are responsible for safety and security. The federal regulatory system compiles information on all emergency transportation events that have occurred in various territories in Russia and determines the preparedness of each territory to handle safety and security situations. There are 7,000 safety inspectors to assess preparedness.

Civil aviation provides a good example of international cooperation for mass transit based on the guidelines of the International Civil Aviation Organization, a specialized agency of the United Nations. International agency safety inspections are to be carried out in all of the 193 participating countries. To date, 150 countries have been subjected to inspections. Russia conducts joint inspections (announced and unannounced) with the U.S. Transportation Security Administration at airports in Russia and the United States. The most valuable experience has often been these direct exchanges between member countries.

In his presentation on counterterrorism awareness training for mass transit, Joseph Bober of the New Jersey Transit Police Department pointed out that since September 11, 2001, there has been a shift from traditional policing methods to a homeland security outlook. Since it is not possible for police to patrol all modes of transport at all times, technological and human resources are used as force multipliers. Technological resources include closed-circuit television, card access, intrusion detection, and interoperable communication systems, the latter being particularly important. Human resources that assist in policing are the partnerships and associated training involving 560 other law enforcement agencies, daily commuters, employees of many facilities, and the general public. Often, the general public observes events that provide significant pieces of information. It is important to emphasize in public awareness training how to report such information to authorities.

Training the community to identify and report possible threats is paramount to countering terrorism. Community policing efforts include training transit police officers in behavioral assessment, improvised explosive device recognition, and counterfeit identification recognition. Employees of many components of the transit system attend training programs in system security awareness and training programs for community emergency response teams. There is also a community outreach informational program targeted on commuters and the general public. The use of the media to advise citizens on appropriate actions during a crisis is a particularly challenging task due to the potential for misuse.

Vladimir Lopatin of the Research Institute of Intellectual Property discussed national and international priorities in countering transportation terrorism. Seventy percent of terrorist attacks rely on various means of public transport to convey the terrorists. However, terrorism threats in transit were not identified as a priority concern until recently.

He emphasized the role of scientists in fighting terrorism. An analysis of transportation legislation and law enforcement in 2000-2001 indicated that a strong antiterrorism component was missing from transportation policy. In 2001, scientists recognized the need for interaction between government, science organizations, and the business community.

In May 2002 a high-level advisory group was established in Russia to address counterterrorism measures in the transportation sector. To date, scientists and practitioners on terrorism and transport security have convened at six international conferences to share lessons learned based on experiences, with their findings and insights being published in the form of proceedings.

Lopatin listed several priorities for international cooperation:

- Balancing the security of the general public with their civil rights
- Transferring security experience from one mode of transport to another, such as from aviation to rail transport
- Assuring continuity of effective counterterrorism and security methods
- Developing and implementing common standards for security
- Continuing scientific assessments of counterterrorism methods that should be considered in preparation of government regulations and in carrying out nongovernmental activities in the transport sector

Mordecai Dzikansky, who is affiliated with the New York City Police Department, presented a strategic approach to protecting transportation facilities, drawing on his experience as an overseas liaison representative to the Israel National Police (INP). Detective Dzikansky described the INP's approach. After an attack, the affected area is divided into three zones: (1) the inner zone, the scene of the attack; (2) the second zone, where forces are gathered, victims are treated, and the joint command center for the first responders is located (spectators are evacuated from this zone as well); and (3) the third zone, which extends to the outermost limits of a sealed area, using roadblocks to contain any potential suspects. Establishing these zones and understanding the roles and responsibilities of the various teams responding to the event allow the scene to be evacuated, secured, evaluated, and cleaned up in an efficient manner and time frame.

Although the number of terrorist attacks of all types carried out in Israel has decreased since 2002, the number of attempted attacks has significantly increased. Suicide bombings are still the simplest and most effective method to maximize casualties. They often involve components of transportation systems.

Consequently, it is important to recognize suspicious persons based on their appearance and behavior. Suspicious appearances can include inappropriate clothing for the season, place, or time; luggage that is incompatible with the surroundings; protrusions in clothing; and concealment of the hands. Indicators of suspicious behavior include excessive nervousness, profuse sweating, walking slowly while focusing on sidewalks, a determined walk, and an inability to carry on a coherent conversation when questioned.

The presentation concluded with a discussion of the various methods Israel employs for defending against terrorist attacks. They include controlling boundaries, utilizing highly professional security guards, training the community to be vigilant, and employing roadblocks. Of utmost importance is high-quality security intelligence.

Vladimir Cherepenin of the Institute for Radio Engineering and Electronics described the application of magnetic inductive tomography for control of passenger flow. His presentation focused on magnetic inductive tomography as a tool for a screening portal. This type of passenger screening raises concerns over invasion of privacy, however. In any event, it is still in the prototype stage of development and will require considerable study to resolve the privacy issue.

Adolf Mishuev of the Blast Resistance Research and Development Center discussed measures and technologies for ensuring blast proofing and blast resistance for transportation, industrial, energy, and civil facilities. Of course, fires occur much more frequently than blasts—perhaps 500 times more frequently. Thus, most studies emphasize flame resistance, and a new goal is to minimize damage from terrorism explosions.

Mishuev focused primarily on tunnels. One example of an antiterrorism device is the gas analyzer, which can be installed in a tunnel and, if an unusual gas reading occurs, can send a signal to a traffic light, stopping traffic before detonation occurs. Explosion of 1 kilogram of some gases (for example, acetylene or methane) can be roughly equal to 10 kilograms of TNT. Gas does not detonate; it conflagrates.

In his presentation on transportation planning for evacuations, John Falcocchio of Polytechnic University discussed the importance of an integrative evacuation process involving transportation agencies; first responders; and local, state, and federal government agencies. Technology was identified as an important component of the evacuation process—informing travelers of circumstances clearly, reliably, and in a timely manner; monitoring the impacted area for real-time observation; and controlling the movement of vehicles and persons in the transportation system. Training programs and exercises are important to evaluate evacuation plans at various phases of preparation, response, and recovery.

Viktor Dosenko of the Business Development Fund at the International Congress of Industrialists and Entrepreneurs and Gennady Taranenko, advisor on science and security at the International Academy of Transport, made a joint presentation. Their focus was on organizing the operation of complex transportation systems. The discussion raised the issue of international transportation standards that are most appropriate, including, for example, construction standards for facilities and electrical standards for operations.

George Bugliarello of Polytechnic University discussed interfacial vulnerabilities of transportation systems. Of special interest is the lack of attention to the interfaces between biological, social, and machine components (BIO-SO-MA) of transportation systems. Examples of the biological component are the operators, security personnel, and individual users. Social components include an organization's operating entities, government, and the community. Machine components are vehicles, platforms, access facilities, power and fuel supply, and communication networks. Inadequate attention to the interfaces can often make an attack more effective. The vulnerabilities of the interfaces can take many forms; for example, the BIO-SO interface depends on individuals and organizations taking the initiative to alert interested parties. The BIO-MA interface may reflect inadequate operational knowledge, machine failure, or sabotage. The response to Hurricane Katrina revealed failures in all three interfaces.

Nikolai Makhutov, Vitaly Petrov, and Dmitry Reznikov of the Institute of Machine Sciences made a final presentation on specific features of terrorism with an emphasis on analysis of risk and damage for natural and technogenic catastrophes of various magnitude and frequency.

SITE VISITS

Research Institute for Civil Defense and Disaster Management

The visit to the Research Institute for Civil Defense and Disaster Management began with a discussion of the role of the institute in supporting decision making at the regional level for protection of the population and territories during emergency situations, including those caused by terrorist acts. The institute was established in 1976. Recently, its staff has done considerable research on the consequences of terrorist attacks. They model emergency situations using algorithms to help improve response scenarios.

A monitoring center has several workstations that process incident-related information, gather geographical information, assess tasks for appropriate responses, and transmit information to first responders. Monitoring and response depend on effective interaction between systems, such as sensor systems, a system for managing emergency resources, ventilation systems, light and sound warning systems for the public, electrical systems for support of buildings, and video surveillance systems. Requiring that these systems be interconnected improves the chances of preventing and eliminating emergency situations. Emergency packs are available for sale to members of the public.

The institute has algorithms for simulating and combating terrorist attacks in the subway system. The focus is on the biological factor, and particularly the human factor. In the first stage, the attack must be recognized as a terrorist attack. Algorithms facilitate assessments of the validity of incoming information. At times, information comes from many sources and can be contradictory. In the second stage, the human and technological capabilities needed to respond to the attack are identified.

A national crisis management center is being set up in Moscow to oversee coordination of the various agencies responding to an attack. At the subnational (city) level, monitoring system centers are being set up to receive information in real time. With preset algorithms in place, decisions can be made quickly to determine necessary actions.

Sergei Todoseichuk described equipment and technology for emergency response and prevention activities. Some examples are airplanes capable of carrying and dispersing 12 tons of water for fighting fires; robots for detection and handling of chemicals, biological agents, bombs, and radioactive debris; vehicles for transporting response teams; and mobile stations to temporarily replace infrastructure damaged during an emergency, including telecommunications facilities. He also described approaches for evacuations from high-rise buildings, such as helicopter landing sites on rooftops, devices installed in new buildings to assist individuals in descending during an emergency, and emergency response ladders capable of reaching the 40th floor of a building.

Valery Akimov of the Center for Strategic Research of EMERCOM presented an analysis of Russian rescue service activities in the elimination of consequences of terrorist acts in cities and on transportation in recent years in Russia and adjacent countries. At this time, a main weapon deployed by terrorists is vehicles, but unmanned aircraft may soon be used. Unfortunately, the coordination of agency interests may be difficult. For example, EMERCOM's objective is to save lives, while other agencies are charged with eliminating the terrorist threat. This contradiction of objectives was illustrated during the October 2002 hostage incident in the Dubrovka Theater in Moscow. There was a lack of communication regarding the type of gas to be used by Russia's Federal Security Service (FSB) in neutralizing the terrorists. EMERCOM was not prepared to protect personnel from the gas, and this deficiency reduced the success of the rescue effort. This case study was discussed in detail at a previous interacademy workshop.[1]

Aleksei Popov of the Center for Information Technology of EMERCOM discussed the development of automated systems for a single emergency dispatch service. The current system relies on separate emergency dispatchers for fire, police, and medical services. A unified duty dispatch service for fire, police, medical, emergency, gas leak, and antiterrorism response is in development, with a single emergency number (112) to be set up in 2008. A national crisis center is being established for the dispatch service for 112 calls. The 112 calls received by the regional dispatch service will be reported to the unified dispatcher.

Research Institute for Fire Protection

The Research Institute for Fire Protection is part of the State Fire Service and is the main fire engineering research center in Russia. The institute participates in research and in implementation of state scientific and technological policy in the field of fire safety. The institute maintains extensive information on fire emergency situations, regularly analyzes the information, and provides support for the implementation of management decisions. The institute has a situation center that conducts mathematical modeling of fires.

Modeling helps determine the number of firefighting crews needed for a given incident and where to deploy them. The total time taken to detect a fire, receive the information, alert the fire brigade, dispatch the fire brigade to the site, and extinguish the fire should all be less than the time it takes to evacuate a burning building. The firefighting system is being reformed and will be divided into five divisions: (1) federal, (2) subnational, (3) sectoral (government ministries and departments), (4) municipal, and (5) privately owned and volunteer.

The site visit concluded with a tour of the building where various pieces of equipment are tested to determine whether they meet fire code standards and where new materials and methods for fire safety are also tested.

NOTE

1. Kolesnikov, Y. 2004. Lessons learned from the *Nord-Ost* terrorist attack in Moscow from the standpoint of Russian security and law enforcement agencies. Pp. 26-34 in Terrorism: Reducing Vulnerabilities and Improving Responses: U.S.-Russian Workshop Proceedings. Washington, D.C.: The National Academies Press.

3

U.S.-Russian Working Group on Energy System Vulnerabilities

A. Chelsea Sharber (Rapporteur)

The Working Group on Energy System Vulnerabilities met March 19-20, 2007. The Nuclear Safety Institute (IBRAE) of the Russian Academy of Sciences (RAS) served as host for the meetings. Working group members made presentations on a wide range of issues concerning energy systems. The U.S. participants also made site visits to the Central Gas Control Department of Gazprom and the Rosenergoatom Crisis Center.

WORKING GROUP PRESENTATIONS

Ashot Sarkisov of IBRAE described the development of a strategic master plan (SMP) as an example of an approach to making decisions on global safety and security issues. Russia developed an SMP for decommissioning retired nuclear submarines and surface vessels and for carrying out environmental rehabilitation after a mass retirement of ships in the late 1980s and early 1990s. At that time it was necessary to reduce the number of nuclear-powered vessels in operation.

The SMP calls for an integrated approach to vessel disposal involving many organizations, a variety of laws and regulations, and many technological approaches. The SMP deals not only with naval submarines but also with icebreakers owned by private companies. Internationally, the Northern Environmental

Partnership was established to address environmental problems, including nuclear submarine and icebreaker disposal and also nonnuclear environmental problems. It accepts funds from international donors, and in these cases the emphasis is of course on joint planning.

Before the SMP was developed, there was no properly justified concept for coping with problems of submarine decommissioning and environmental rehabilitation, and the use of available financial resources to solve the problems was far from optimal. The SMP has been very important in improving the situation.

The SMP does not include a strict short-term time line. However, it adheres closely to the dual objectives of prompt disposal of nuclear vessels and preservation of the environment. The SMP integrates all previously developed planning aspects that had been approved. Concepts for some parts of the SMP were not well defined, and a series of strategic studies has been undertaken to elaborate these concepts. More than 70 high-priority projects are under way pursuant to the more than 40 approaches justified within the framework of the SMP.

Sergei Serebryakov of the RAS Oil and Gas Research Institute led a discussion of pipeline security. Russia has 34 percent of proven world reserves of natural gas and 13 percent of world oil reserves. Natural gas is shipped through a series of Gazprom trunk pipelines. System operation stability is achieved by effective reliance on diagnostics and timely repair. Today the system operates at nearly 100 percent capacity. There are plans for further expansion of the system by Gazprom and independent companies.

The 2005 report of the Federal Authority for Industrial Safety Issues identified the following pipeline hazard factors:

- Stress corrosion (for pipelines built more than 15 years ago)
- Theft from pipelines
- Accidents due to poor quality installation practices and assembly work and poor quality assurance

To help ensure industrial safety and security, owners such as Gazprom, Transneft, and Transneftprodukt have approved plans for reconstruction and major overhaul of selected facilities. Also, all entities have plans for countering terrorism. To date, there have been several acts of terrorism carried out against pipelines in Russia (for example, in Dagestan).

Pipelines and trunk lines make attractive targets for terrorists worldwide, as has been seen recently in Iraq. Serebryakov contended that the political situation in Iraq has prompted attacks against pipelines in Sudan, India, Turkey, Colombia, and Nigeria. He added that the pessimistic conclusion is that as long as gas and oil are the basis for economies in many areas of the world, it is difficult to eliminate the terrorist threat completely.

Vyacheslav Kuznetsov of the Russian Research Center—Kurchatov Institute described underwater technologies for transportation of liquefied natural gas

(LNG) and the strengthening of global energy safety. An efficient, reliable, and safe energy supply can be obtained by reviewing vulnerabilities of local energy systems; detecting weak points; and developing international cooperation efforts, shared technologies, and adequate protection. Some steps are contingent on the globalization of the gas market, however. There have been no terrorist attacks to date on LNG tankers, but there have been accidents. LNG production growth currently exceeds natural gas production, and there are plans to expand production. At the same time, LNG facilities also provide a high-impact target for terrorists. With subsurface transport and careful management of LNG, the risk of terrorism would be greatly reduced, creating a better economic and technical situation, according to Kuznetsov.

Yury Parfyonov of the RAS Scientific Association for High Temperatures addressed electromagnetic terrorism and the threat to a nation's energy infrastructure. Electromagnetic terrorism involves the use of strong electromagnetic pulse (EMP) transmitters and high-voltage pulse generators that can damage flight control systems, telecommunication systems, electromagnetic devices at nuclear power plants, information systems, technical systems of environmentally hazardous facilities, and electrical power generating facilities. Several examples of small EMP devices were described. Research has focused on the effect of EMP devices on various kinds of systems. Facility designers must take all possible measures to protect electronic systems. Some of the suggested measures for protection include international cooperation, joint experiments on the topic, and both Russian and international standards for use of the technology.

Siegfried Hecker of Stanford University made a presentation on industry-sponsored studies of the vulnerabilities of U.S. power systems. There have been numerous energy vulnerability studies in the United States, including studies by the Electric Power Research Institute (EPRI) and the U.S. Energy Association. The EPRI studies focus on the nature, consequences, and mitigation of a terrorist threat. A terrorist attack could be perpetrated on the energy grid itself, could use the energy grid as an attack medium (dispersion of chemical or biological agents through natural gas pipelines), or could use the grid to amplify the attack (for example, through EMP). The energy grid in the United States is vulnerable because it is centralized, its complexity continues to increase, communication on the grid is not secure (sometimes occurring via the Internet), supply and transmission nodes are easily accessed, energy companies are not aware of their own vulnerabilities, and the response to an attack is insufficient or poorly coordinated.

Strategies to counter disruption include greater government efforts to prevent attacks, more resistant facilities, prompt restoration of damaged facilities, and new capacity additions for the energy system. Short-term protection measures are multifaceted. They include, for example, ensuring that Internet connections are secure, checking gas pipelines and electrical grids through drones, and preparing probabilistic vulnerability assessments of the physical infrastructure as a basis for identifying weak spots. Installing more natural gas leak detectors and protection-critical substation components are obvious steps that can be taken. Preventing dispersion of carbon fiber and Mylar chaff may also be important in some situations.

Medium-term protective measures include breakaway devices that prevent a cascade and line breakers that switch and reroute when necessary. A secure and private wide-area communication network with backup can be critically important. As final examples, more natural gas storage capacity is needed, and more efficient electric driver compressors to switch the direction of gas flow are important.

In conclusion, Hecker noted that in the policy area, little has changed in the United States in the past 20 years. Federal energy policy continues to emphasize reliance on options with significant vulnerabilities as compared to alternatives. Policy tends to ignore or at least minimize many resilient options that can make the system efficient, diverse, and dispersed.

Siegfried Hecker also made a presentation on the security of nuclear power plants on behalf of John Ahearne of Sigma Xi, who had prepared the presentation but could not participate in the meetings in Moscow. As is well known, the U.S. Nuclear Regulatory Commission (NRC) is responsible for nuclear power plant security. After September 11, 2001, nuclear power plant security guidelines were revised. The NRC also specified requirements for training, access, security officer working hours, defense strategies, mitigating measures, and integrated response.

Recent changes dealing with security of nuclear power plants are detailed in the Energy Policy Act of 2005. They include background checks for personnel with access to any weapons, as well as access authorization programs for personnel who use computer systems affecting operation safety, security, and emergency response capabilities. Other recent changes designed to safeguard information include increased security of personnel, increased patrols, increased physical barriers, vehicle checks farther from facility entry points, improved coordination with the military forces, better security and emergency response training, and restricted site access.

After September 11, 2001, there has also been an increased focus on security of radioactive materials and spent nuclear fuel. Aircraft attacks on nuclear power plants are also a concern. Design criteria for new reactors should include the capability to withstand an aircraft attack. Finally, a lack of adequate understanding of technological performance, suppression of information on security violations, and lack of engineering recognition of the fragility of some systems sometimes create obstacles to objective analyses.

Drew Lieb of the New Jersey State Police discussed homeland security concerns about energy facilities. New Jersey is the most densely populated state in the United States, located between Philadelphia and New York City, and contains many industrial facilities, including four nuclear power plants and a proposed new LNG terminal. Northern New Jersey is home to oil companies, chemical plants, energy facilities, and transportation hubs. Southern New Jersey houses the state's major concentration of oil refineries.

Hence, New Jersey provides attractive targets for terrorists. Several terrorist attacks have been planned or attempted from within New Jersey. As a result, statewide coordination is necessary. A homeland security department consisting of two parts (emergency management and special operations) has been established to protect the state.

The Emergency Management section works in parallel to the national Federal Emergency Management Agency and handles natural, industrial, and nuclear disasters and terrorist attacks. It organized the Top Officials (TOPOFF) 3 simulated terrorism attack exercise in coordination with the state of Connecticut and the United Kingdom. It works with private security personnel of the nuclear facilities in New Jersey on training and the Regional Operations Intelligence Center on security. The Emergency Management section is also involved with pipeline security and the consequences of catastrophic events, such as the 2003 Northeast power blackout.

The Special Operations section includes the following divisions: hazardous materials, canine, aviation, marine services, government security, arson/bomb, and transportation security. They are all involved in protection of the state's infrastructure. The Marine Services Bureau protects the state's nuclear power plants, which are located along the coast.

Boris Krupchatnikov of the Nuclear Industrial Environmental Regulatory Authority (Rostekhnadzor) made a presentation on current Russian requirements regarding protection of nuclear power facilities. During the history of nuclear power generation in Russia, the focus has shifted back and forth between safety and security. The Federal Law on Atomic Energy stresses that physical protection must be provided in all stages of power generation. Nuclear operators must provide physical protection, with one exception. If an operator cannot guarantee physical protection, the responsibility is shifted to state authorities.

Regarding supervision and compliance to regulations (licensing requirements), the function of Rostekhnadzor is similar to the NRC. The four levels of regulatory requirements are (1) federal law, (2) government acts, (3) Rostekhnadzor acts, and (4) agency regulatory documents. Physical protection requirements are based on International Atomic Energy Agency recommendations, and inspection efforts are based on internationally accepted principles.

John O'Neil of the U.S. Department of Homeland Security (DHS) discussed DHS science and technology interests in countering terrorism in 2007. A major realignment of effort in the DHS Directorate for Science and Technology was under way, including a framework for a customer-focused, output-oriented science and technology management organization and the possibility of science and technology liaisons embedded worldwide. This step should provide a better basis for international science and technology collaboration. At that time, the Directorate for Science and Technology had six divisions: (1) explosives; (2) biological and chemical; (3) command, control, and interoperability; (4) borders/maritime; (5) human factors; and (6) infrastructure and geophysics. These divisions supported research, including work with universities, DHS centers of excellence, U.S. Department of Energy laboratories, and DHS laboratories.

Two models to support research were being established: (1) an industry board-of-directors model with the customer defining the need and (2) the consensus process model with national capability gaps defining the need. Both were unusual models for action in the U.S. government and highlight the importance of the realignment that was under way.

Raphael Perl of the Organization for Security and Cooperation in Europe gave a presentation on a strategic approach to protecting energy facilities. He emphasized that it has not yet been decided how to deal with terrorist attacks on the energy sector in the United States. When terrorists attack the infrastructure, what do they want to accomplish? When governments respond to attacks, what do they want to achieve? Six policy issues should be considered:

1. How can critical links in infrastructure be reestablished without rebuilding the entire system? What network disruptions cause the worst effects? What causes an attack to have national impact?
2. How is the threat assessed?
3. What and whose information is used in threat assessment?
4. How are potential consequences assessed?
5. How is risk reduction best accomplished?
6. How are resource requirements prioritized within the framework of a master plan?

Two conceptual approaches to the possibility of terrorist attacks are important: (1) security and strengthening of the infrastructure and (2) infrastructure recovery if attacked. Will the capability for a quick recovery make us a less attractive target for terrorists? Don't the goals of terrorists go beyond infrastructure damage to include economic damage and paralysis, confusion, lawlessness, and loss of confidence in the government? And should this expanded set of goals be taken into consideration when preparing to respond?

Vitaly Gridin of the RAS Oil and Gas Research Institute discussed the safety of gas pipelines. He described a theory based on biorhythmic cycles and the intervals in which geodynamic hazards are more likely. Satellite technology and statistics have been utilized to identify zones with geodynamic shortcomings where cave-ins, faults, and so forth, are more likely to occur, with examples cited involving accidents at mines and oil wells.

SITE VISITS

Central Gas Control Department of Gazprom

Anatoly Paramonov, deputy head of the department, emphasized that Gazprom provides the entire nation's gas supply. With any impact on the system, whether a natural disaster or technical event, Gazprom attempts to mitigate consequences and maintain production capacity. After notification of such an event, an entity other than Gazprom categorizes the problem. Gazprom is not an emergency response center. Special communications channels exist and are used when an incident occurs. Gazprom is simply notified that there was an event, in accordance with the list of persons and entities to be notified for appropriate response, with the goal that consumers should not feel the consequences of such an event.

According to Paramonov, consequences can usually be isolated despite the Soviet heritage of an integrated grid and flow of power, despite the large regions, despite the large number of time zones, and so forth. If there is not enough gas for some consumers, a switch can be made to oil or coal. Gazprom is obligated to follow such an order.

Russian gas pipelines are now connected to Europe.

Gazprom's task is to maintain a balance of resources and distribution. Gazprom has nearly real-time information available on pipelines, with an 8-minute delay. However, many factors can impact the task of ensuring distribution, such as weather, the human factor, and so forth.

Rosenergoatom Crisis Center

Igor Gorelov, head of the Rosenergoatom Crisis Center, and his deputy, Boris Pivnenko, provided a brief history of the Crisis Center. In 1987, after the Chernobyl accident, a government decree was issued regarding the safety of nuclear power plant operations. A questionnaire was prepared to determine what was needed. From architects and designers, from the Ministry of Defense and the Ministry of Health, and from many other organizations, suggestions were offered concerning how information was processed and dispatched at control centers. Materials from the Three Mile Island accident as well as Chernobyl were studied.

After a search for an entity within the country that had addressed closely related problems, the Kaliningrad Space Flight Command Center, with its rapid-response capabilities, was chosen as a model, and a contract was initiated with that center. There was great interest in the approach for involving experts in unusual situations that had been developed at the Space Flight Command Center. The center cannot control flights, and similarly the Rosenergoatom Crisis Center cannot control nuclear power plants.

Then in 1992, after many studies of the Russian experience, a 7-year cooperation agreement was initiated with Électricité de France to develop operations documentation covering emergency procedures in the nuclear power sector.

The Rosenergoatom Crisis Center system monitors online all power units in Russia. The Crisis Center is supported by 11 other centers, connected via videoconferencing capabilities to the Crisis Center. All 11 centers operate 24 hours a day. Ten centers receive real-time information, and the 11th is being upgraded to have this capability. The Crisis Center receives a summary of the activities of 21 nuclear complexes operated locally. In an emergency, the center can access these complexes' information. One lesson drawn from the Chernobyl accident was the lack of coordination of efforts during a crisis. Now there are annual drills, often observed by invited representatives of other countries. Antiterrorism drills are prepared by other agencies, and the Crisis Center participates in these drills. All emergency technical support centers are located a short distance from the Crisis Center, and arrangements are in place for rapid transportation between the centers as needed. There is a dedicated, secure, non-Internet channel for communications. In an emergency, one person from each center comes to the Crisis Center. The Crisis Center has operated with its current level of technical capabilities for 3 years.

Selected Papers

4

Tendencies in Global Terrorism

Raphael Perl,
Action Against Terrorism Unit,
Organization for Security and Cooperation in Europe

I will begin with a general discussion of trends in terrorism. I will then highlight some trends identified by the U.S. government, drawing heavily on two documents: (1) the Department of State's latest version of *Country Reports on Terrorism*, which covers reports for 2005,[1] and (2) an unclassified version of an April 2006 U.S. National Intelligence Estimate (NIE), which includes key judgments relating to terrorist activity.[2] I will conclude with some personal observations on these evolving tendencies.

DEFINING TERRORIST TENDENCIES

Trends in terrorism can be defined as changes in incidents, attitudes, and other factors over time. Trends can be important indicators of levels and types of terrorist activity, can help governments formulate responsive counterterrorism strategies, and can assist both policy makers and policy implementers in allocating resources effectively.

Standing alone, a trend is not necessarily good or bad. It depends on the outcome. For example, a trend by terrorist groups to focus on megaterrorist events might result in an overall decrease in casualties from smaller acts of terrorism over an extended period of time. If measures to counter or defend against such mega-events prove effective, the net result is a decrease in casualties.

Of bottom-line importance is whether the overall *momentum* of terrorist activity is growing or declining. A relevant issue is the degree to which government bureaucratic institutions can work smoothly together and stay ahead of the methods utilized by individual terrorists and terrorist networks. Important as well is improvement in recovery capabilities of states following terrorist acts.

Another pertinent factor is the growth or decline of phenomena perceived by terrorists as directly related to advancing their cause or detracting from it, such as the number of governments that embrace appeasement policies and the amount of media coverage their groups receive. A related issue is how the policies of governments such as the United States and Russia affect popular support for and recruiting by terrorists.

Governments need to collect meaningful trend data, even if the data are unfavorable toward them. As the global economic, political, and technological landscape evolves, and as terrorists seek to surprise and attack the enemy through more fluid organizational structures and new innovative approaches, the nature of the data being collected needs to change. A major challenge facing the counterterrorism community is the need to facilitate acquisition and incorporation of new data indicative of trends while maintaining the continuity of earlier findings.

Trends in terrorism are often shaped by trends in counterterrorism. One trend frequently cited in the media is the decentralization of al Qaeda, which is arguably the result of aggressive U.S. targeting of the organization, its leadership, and its command-and-control capabilities. Hardening U.S. government physical infrastructure overseas or at home might encourage terrorists to shift the focus of attacks to softer nongovernment targets. Similarly, government implementation of better systems to evaluate the authenticity and content of travel documents at lawful ports of entry might prompt terrorists to switch from the legal entry tactics employed by the September 11, 2001, hijackers to illegal border crossings at unsecured locations. Familiarity with future U.S. counterterrorism strategy and tactics and the strategies of other nations is therefore essential for predicting and understanding future terrorist responses.

SIGNIFICANCE OF TERRORISM TENDENCIES

Understanding trends in terrorist activity can assist policy makers in several areas, including (1) better protecting the nation against terrorist attacks, (2) better targeting terrorists and terrorist activity, (3) better prioritizing antiterror resources, and (4) showing antiterror progress when it has been achieved.

It is natural to assume that decreases in terrorist activity, or even slowing the rate of increase, reflect progress in antiterror efforts. However, this type of measurement may underestimate the varied and multidimensional nature of terrorist actions. The often asymmetric, nonlinear nature of terrorist operations, frequently characterized by abrupt changes, increases the deadliness of the threat and may necessitate more comprehensive measurements of trends to reflect this additional danger more accurately.

A common pitfall of governments seeking to identify or enumerate trends is overreliance on *quantitative* indicators at the expense of their *qualitative* significance. One qualitatively creative incident may immediately prove to be a trend by sparking copycat follow-on terror incidents with a resultant change in terrorists' strategy, tactics, and targets.

TENDENCIES IDENTIFIED BY THE DEPARTMENT OF STATE

Each year, the Department of State produces an annual report on terrorism that is considered by many to be one of the best analyses of global terrorist activity. The *Country Reports on Terrorism 2005* and the underlying data portray a threat from radical jihadists that is becoming more widespread, diffuse, and deadly and increasingly homegrown. This phenomenon of looser, more local networks was manifest in the July 2005 terrorist attacks on London's transit system, in attacks the previous year on trains in Madrid, and more recently in the June 2006 arrests of 12 men and 5 juveniles in Ontario. The current report concentrates on terrorist activity for 2005, and trends for 2006 are expected to parallel those identified in this talk.

Three trends are identified by the Department of State as follows:

- **Microactors** are a new phenomenon. This development is spurred by perceived U.S. and allied successes in isolating and killing much of al Qaeda's centralized leadership, thereby reducing its centralized command-and-control capability. The result is an al Qaeda that is assuming more of an ideological and propaganda role rather than an operational role, with the operational component of the movement increasingly being assumed by small autonomous cells and individuals, often homegrown. Such operatives are likely to be technologically savvy, and because they are new to the terrorism landscape, their decentralized actions can be extremely difficult to detect or counter. A logical outcome from such a development is likely to be a growing number of microactors in future terrorist attacks, particularly those involving conventional bombs and bullets. These microactors are relatively unseasoned and unskilled in terrorism tradecraft.
- **Sophistication** is the second trend. Increasingly, terrorists are exploiting the global interchange of information, finance, and ideas to their benefit. They are also improving their technological sophistication across many areas of operational planning, communications, targeting, and propaganda. The effective worldwide orchestration of a campaign against publication by a Danish newspaper of cartoons degrading the Prophet Mohammed is an example of such sophistication.
- **Overlap with international crime** is a third trend reflected in the report. Such a trend, to the extent that terrorists do indeed use the same networks used

by criminal groups, creates a major vulnerability. The more terrorists engage in nonterror forms of criminal activity, the more likely they are to show up on the law enforcement radar screen.

Also cited in the report is an increase in suicide bombings,[3] as well as a strong connection between Iraq and the broader war on terrorism. Terror incidents in Iraq, according to the report, accounted for almost one-third of all terror incidents in 2005 and more than one-half of all terror-related deaths worldwide. Moreover, there is concern among many that Iraq will become an exporter of seasoned terrorists, weapons, and tactics, especially the use of improvised explosive devices (IEDs), with "spillover" not only into the neighboring Gulf region but also into Europe and other regions. Of course, it is interesting to consider the trend data without including the special case of Iraq.

In short, the State Department's *Country Reports on Terrorism 2005* supports the contention that the threat from small terrorist groups or lone terrorists is rising, as is the potential for such microactors to inflict deadly harm and costly economic damage. Such a trend, according to the report, could mean that the immediate future will bring "a larger number of smaller attacks, less meticulously planned, and local rather than transnational in scope."

TENDENCIES IDENTIFIED IN THE APRIL 2006 U.S. NATIONAL INTELLIGENCE ESTIMATE

National Intelligence Estimates are widely considered to reflect the collective judgment of the U.S. intelligence community at the time of their compilation. An unclassified version of key judgments from an April 2006 NIE relating to trends in global terrorism identifies what can be broadly characterized as 10 basic tendencies in global terrorism. The tendencies are as follows:

1. The number of jihadists is increasing worldwide, in terms of both numbers and geographic dispersion.
2. We will continue to see an increasing number of terrorist attacks against the United States and U.S. interests worldwide.
3. The threat from self-radicalized terrorists and groups will likely increase.
4. Terrorists will increasingly employ IEDs, widely used in Iraq, on soft targets outside Iraq.
5. Fighters trained in Iraq will likely provide leadership to terrorist groups outside Iraq.
6. Jihadist groups will continue to seek chemical, biological, radiological, and nuclear capabilities.
7. A rise in radical ideologies (other than jihadist) can be expected. It is anticipated that to some degree this rise will be rooted in anti-U.S. and antiglo-

balization sentiment and that such groups may increasingly resort to terrorist tactics.

8. Terrorist groups will increasingly rely on the Internet to communicate, propagandize, recruit, and train adherents and to obtain logistical and financial support.

9. Europe will remain an important venue for recruitment and staging of terrorist attacks as well as a key target for terrorist attacks.

10. We can expect an overall ongoing trend towards urban terrorist attacks, which relates directly to the topic of our meeting today.[4]

A PERSONAL ASSESSMENT OF BASIC TERRORIST TENDENCIES

I would now like to provide some personal thoughts on the evolution of basic terrorist tendencies. Some of my views have been incorporated or are in the process of being incorporated into U.S. government assessments. Others, however, are not.

- Generally, the use of terrorism as a tactic is becoming more frequent, more geographically widespread, and more deadly. Statistical data provided by the National Counterterrorism Center (NCTC) indicate steady annual increases in the number of terrorist incidents in a growing number of locations. At the same time, NCTC data indicate that a smaller number of incidents are resulting in a higher number of persons killed or injured.
- Terrorism is becoming more indiscriminate in its choice of victims. Casualties include not just combatants, Westerners, and non-Muslims. Targeting is less directed to specific individuals.
- Muslims are increasingly becoming victims of jihadist terrorism. This tendency will increase as Muslim versus Muslim conflict spills over into Europe.
- Terrorism is becoming more focused on economic targets and causing economic damage. The energy infrastructure will increasingly become a target, and the financial infrastructure may well follow. If our economic system cannot deliver concrete benefits to the world's masses, the pool of angry and dissatisfied masses will grow, and terrorists will continue to hijack this dissatisfaction and channel and manipulate it for their causes.
- Jihadist terrorism is increasingly becoming an equal opportunity movement. Groups are actively recruiting non-Muslims, women, and youth.
- Terrorism is becoming more multidimensional. Groups like Hamas and Hezbollah have long had political, social, and religious, as well as military, components. Such groups fill important social service vacuums, obtaining popular support.
- Increasingly, we see a blurring between terrorism and organized crime. This means the emergence of more hybrid organizations. At some point, this may

result in a change to the definition of terrorism where the defining characteristic will be the tactic used and not the motivation behind its use.

- Terrorist groups will increasingly rely on each other for logistical support. This phenomenon has become more widespread in India.
- Explosive devices will likely be used as dispersion mechanisms for chemical agents to be spread beyond Israel to Iraq, Europe, and elsewhere. Currently, groups such as Hezbollah and Hamas add rat poison, a powerful anticoagulant, to conventional explosives for the purpose of increasing casualties. It will likely not be long before other groups "piggyback" on this innovative tactic with other, perhaps more lethal, forms of chemical agents.
- I expect we will see an increase in well-orchestrated acts of terrorism against large systems. A significant number of jihadist leaders and key operatives have engineering training, and engineers think in terms of systems and networks. Included here are more multiple attacks, attacks on first responders, and more attacks on urban centers. Increasingly, urban centers are viewed by terrorist groups as a system. Terrorists seek to disrupt the functioning of the system to the maximum extent possible.
- We will continue to see an increase in the number of anonymous terrorist incidents or the number of terrorist incidents that are claimed in a way that masks the true identity of those committing them. This trend is largely rooted in three factors: (1) fear of government response, (2) a desire to amplify fear and mystery surrounding the group, and (3) publicity fading as a goal of some terrorist strikes.
- The phenomenon of self-radicalization and homegrown terrorism will escalate. This will especially be the case in Europe, where Muslim immigrant communities often perceive themselves as subject to discrimination and as not having strong local roots.
- Terrorism will prove to be increasing costly to societies, in both the economic costs of added security and the trade-offs of civil liberties for enhanced security. It is estimated that the increased global macroeconomic costs of added security in the wake of the September 11, 2001, attacks exceed $1 trillion. Moreover, many suggest that the greatest threat posed to societies by terrorism is the threat to the continued existence of democracies with their wide range of freedoms as we now know them.
- I also expect that we will see more terrorism, much more terrorism. As the gap between the "haves" and "have-nots" continues to rise dramatically, terrorists will increasingly sow their intolerance, hatred, and extremism and will recruit from the dissatisfied.

On the other hand, as use of terrorism as a tactic grows and evolves, so does our recognition and understanding of the threat and its basic tendencies. Our experience in containing terrorism without overreacting is growing as well.

NOTES

1. See *www.state.gov/s/ct/rls/crt/c17689.htm*. Note that the *Country Reports on Terrorism* version for 2007, which will cover 2006, is not expected to deviate markedly from its characterization of the terrorist threat for 2005.

2. See Declassified Key Judgments of the National Intelligence Estimate "Trends in Global Terrorism: Implications for the United States," dated April 2006. Available online at *www.dni.gov/press_releases/Declassified_NIE_Key_Judgments.pdf*.

3. Note, however, that although the total number of suicide bombings increased in 2005, it is not fully clear that the ratio of suicide bombings to other forms of attacks has increased concomitantly.

4. Note that the NIE does not specifically identify such an overall trend, but from its overall reading, such a tendency towards urban attacks appears implied and inherent throughout the unclassified summary of the document.

5

Use of Predictive Modeling Packages for Effective Emergency Management*

Nikolai Petrovich Kopylov and Irek Ravilevich Khasanov,
All-Russian Scientific Research Institute for Fire Protection (VNIIPO) of the Russian Ministry for Civil Defense Affairs, Emergencies, and Elimination of Consequences of Natural Disasters (EMERCOM)

INTRODUCTION

About 1,000 major disasters and catastrophes occur each year in Russia. As a result of industrial and other technogenic[1] accidents alone, more than 200,000 people annually are injured or mutilated and more than 50,000 are killed (including traffic accidents). The economic losses from technogenic and natural disasters total 6 to 7 percent of the country's gross domestic product.[2]

An analysis of terrorist acts indicates that providing antiterrorism protection for facilities at risk of fire or explosion is the most urgent and important aspect of guarding against terrorism of a technogenic nature. Given these conditions, the effectiveness of management decisions made in eliminating the consequences of acts of technogenic terrorism largely depends on informational and analytical support and predictions of how fires and emergencies might develop.

The primary goal of the integrated state system for predicting and eliminating the consequences of extreme situations is to integrate the efforts of executive branch agencies at both the federal and the Russian Federation subject levels. The main objectives of activities under the state system are as follows:

---

*Translated from the Russian by Kelly Robbins.

- Monitoring and predicting extreme situations
- Training specialists in emergency prediction and response
- Educating the public on actions to be taken in emergencies
- Developing preventive measures to reduce the risks and lessen the consequences of emergencies
- Improving the management of emergency prediction and response measures

Effectively accomplishing these objectives is impossible without utilizing new information and telecommunications technologies.[3]

The National Crisis Management Center (NCMC) has been created in Russia to unite the information resources and functional capabilities of local subsystems of the integrated state system with the aim of improving the quality and timeliness of management decisions on predicting and eliminating the consequences of emergencies. The NCMC is a geographically distributed information management complex with peripheral elements that make it possible to manage the forces, means, and resources of the integrated state system and civil defense entities during crises and emergencies.[4]

SITUATION MANAGEMENT CENTERS

The rapid development of information technologies has led to the appearance of massive amounts of informational, communications, audio, and video data that must be recognized, structured, and analyzed in order to make competent management decisions. Meanwhile, although the rates of information technology development have increased, the amount of time allotted for making management decisions is being reduced, especially for decisions made in crisis situations.

The strategy for creating and developing national security support systems by states attests to the fact that information and management centers created at the national and regional levels and in major cities represent the universal foundation for the crisis management system. Informational support for such centers is provided by services such as 911, 112, and 01, as well as by scientific and academic centers.[5]

Intensive efforts are under way to apply modern concepts for the creation of crisis management centers involving high-technology equipment for communications and information exchange, depiction, and processing, which helps in efficiently preparing and making well-founded management decisions. Situation centers have been created in Moscow and regional centers in the various territorial agencies of EMERCOM. These centers are complexes of programmatic and technical resources housed in special facilities where emergency response officials may assemble if an emergency arises. The situation centers regularly conduct training exercises, some of which involve members of the commission on extreme situations.

The activities of a situation center represent the most expedient means of implementing the decision support system based on technologies for numerically simulating and creating visual representations of situations and object behavior. They are the top level in the system for managing the organization, the industry, the region, and the country.

The situation center is an information analysis system that makes it possible to assess the real status of the object or event being managed, detect trends as external and internal changes develop, and analyze (simulate) possible consequences of management actions.[6] From the most general standpoint, the situation center (room or hall) could be called a facility from which ongoing emergencies are observed or possible situations are analyzed. However, such an interpretation fails to take many factors into account. The modern understanding of a situation center focuses on the entirety of programmatic and technical resources, scientific and mathematical methods, and engineering solutions for automating processes for situational depiction, numerical simulation, analysis, and management.[7] All of these means and methods make possible the following:

- Providing information on matters where operational decisions are required
- Visually depicting management situations to reveal cause-effect relationships for events being analyzed
- Numerically simulating and conducting situational analyses
- Effecting operational control over efforts being carried out by structural subunits
- Verifying execution of decisions made

The situation centers include various types of analytical support capabilities (programmatic, technical, linguistic, psychological, and so forth). The situation center has four basic levels: (1) scientific-mathematical, (2) engineering, (3) programmatic, and (4) technical.

The scientific-mathematical level includes all scientific theories, methods, algorithms, research, and developments necessary for the activities of the other levels. It provides the foundation for determining the expediency of creating the situation center, defines the effectiveness of its operations, integrates various components, and rectifies errors in a correct and timely manner. The engineering level provides concrete solutions in the selection and development of devices and software. It includes the necessary technological and design calculations, numerical simulations, technical equipment, facilities, program specifications, work algorithms, and so forth. The programmatic and technical levels include the appropriate support necessary for the tasks and functions assigned to the higher levels to be carried out.[8]

The main feature of the situation centers that determines their name is situational (dynamic) simulation. Prediction makes it possible to create scenarios based on analysis of the current situation and existing trends. The situation center allows managers to see newly arising threats in a timely manner and to take measures to counter them.[9]

THE VNIIPO SITUATION CENTER

The VNIIPO Extreme Situation Modeling Center (Situation Center) was established at VNIIPO in 2006 based on the requirements of the Concept for the Creation of the National Crisis Management Center. Functionally, the center is a part of the NCMC. The VNIIPO Situation Center is designed to provide informational, analytical, and expert support for management decisions by officials from operations management agencies in responding to major fires and technogenic emergencies at critically important sites. The center's primary tasks are as follows:

- Collecting, accumulating, and analyzing information on the status of facilities at risk of fire or explosion and on EMERCOM forces, means, and reserves
- Providing informational, analytical, and expert support for management decisions on preventing and eliminating the consequences of fires and technogenic emergencies
- Predicting the development of fires and technogenic emergencies at critically important facilities
- Developing, implementing, and supporting software systems for management and modeling at the Situation Center
- Providing technical documentation for numerical simulation packages for fires and emergencies and organizing and supporting work to develop models and methodologies to facilitate the Situation Center's activities
- Organizing the operations and information security of the Situation Center
- Developing and supporting technical and telecommunications services at the Situation Center and developing and maintaining informational support for data banks and databases

Based on its purpose, functions, and tasks, an organizational-technical structure including the following components has been proposed for the VNIIPO Situation Center:

- Analysis
- Applied software support
- Information infrastructure
- General-purpose software and hardware environment
- Complex of special-purpose software and hardware resources
- Information security subsystem

SUBSYSTEM FOR INFORMATION SUPPORT FOR MANAGEMENT DECISION MAKING

The Situation Center's subsystem for information support for management decision making must be responsible for preparing guidelines and statistical information needed for making command decisions. Databases that have been developed and are being utilized successfully lie at the foundation of the functional complexes and tasks of the information support subsystem at the Situation Center. For example, VNIIPO has created a user version of the informational database "Fire and Explosive Hazards of Substances and Materials and Means of Extinguishing Them" (see Figure 5-1), which is used in more than 100 of EMERCOM's State Fire Service branches. The database contains information on more than 12,000 substances and materials, including data on the fire and explosive hazards of substances and materials, means of extinguishing them, and the potential reactions of substances and materials if they should come into contact.

VNIIPO has developed and is using several regions a geographic information system for decision support in operations management by local fire and rescue units involved in responding to fires and eliminating the consequences of emergency situations. This system provides informational support for the following types of activities:

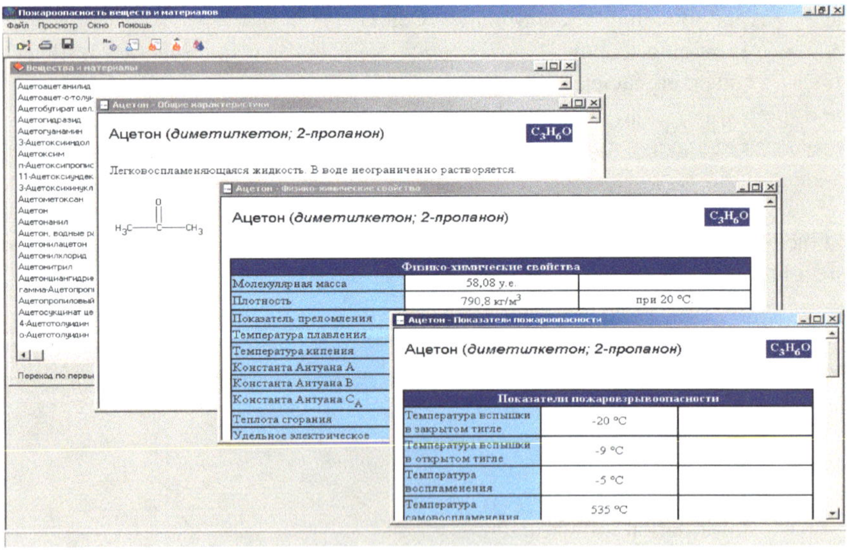

FIGURE 5-1 Screenshot from the database "Fire and Explosive Hazards of Substances and Materials and Means of Extinguishing Them."

- Reception and processing of fire (emergency) calls, including location and formulation of orders for dispatching personnel and equipment to handle them
- Accounting and control of the status and deployment of equipment and weapons
- Redeployment of units, depending on their operating regimes
- Management of operations at the fire (emergency), establishment according to proper procedure of accounting of situation changes and use of personnel and equipment, and registration of necessary information
- Implementation of other measures aimed at ensuring service delivery according to established procedure and increasing the effectiveness of firefighters' actions

An automated decision support system for use by fire captains at the scene has been developed to provide operational information and analytical support for decision makers. This system automates the following processes:

- Accumulation and storage of site data
- Presentation in convenient form of information used by the fire captain in preparing operational decisions on managing firefighters' actions at the scene
- Calculation of potential fire situations
- Calculation of personnel and equipment needed to extinguish fires
- Calculation of delivery systems for means of extinguishing fires, including calculation of pump-hose system parameters
- Preparation of typical command decisions
- Preparation of operational documents
- Creation and correction of databases

SUBSYSTEM OF ANALYTICAL SUPPORT FOR MANAGEMENT DECISIONS

The subsystem for analytical support of management decisions must facilitate numerical simulation and prediction of the development of fires and emergency situations. With the aim of studying major fires at dangerous production facilities or in population centers, a series of studies has been conducted to simulate major fires in open spaces.[10] Based on this research, a numerical simulation has been proposed for the aerodynamics of the environment. It is based on nonstationary Navier-Stokes differential equations, taking into account the effects of turbulence, atmospheric stratification, smoke aerosol diffusion, and phase transitions caused by the presence of moisture in the surrounding air. Figure 5-2 depicts a smoke cloud formed over a fire with a radius of 5 kilometers and a maximum heat transfer of $q_m = 4.7 \cdot 10^4$ W/m^{-2}.

For several decades, the institute has been working to develop and apply

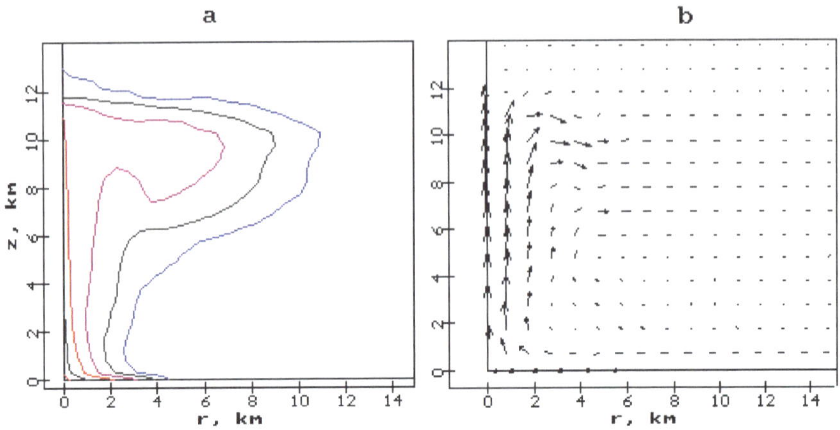

FIGURE 5-2 Isolines of smoke aerosol concentrations (a) and convective flow structure (b) over a fire with a 5-kilometer radius 1 hour after ignition.

mathematical modeling of fires in structures and buildings. Mathematical models are widely used in resolving questions of ensuring the safety of people during fires, designing evacuation paths, and creating fire alarm systems. The various mathematical models of fire development in structures (interior fires) fall into the following three categories:[11]

1. Integral mathematical models (first-generation models)
2. Zone mathematical models (second-generation models)
3. Field (computational fluid dynamics) mathematical models (third-generation models)

Integral fire models are limited to recording physical heat parameters at the level of average values (by volume or by heat-absorbing surfaces).[12] Equations on the development of a fire describe the change in average volume parameters for the situation over time. The system of differential equations for the balance in the structure includes equations on the material and oxygen balance, equations on the balance of combustion products and inert gas, and an energy equation.

An example of the successful use of the integral modeling method would be the study conducted by institute specialists of possible development scenarios for the fire caused by the crash of a Boeing-767 aircraft into the World Trade Center in New York City. Several fire scenarios were considered. The first group of scenarios simulated the combustion of jet fuel spilled from the plane's fuel tanks, the second group of scenarios covered the burning of office furniture, and the third group focused on the combined burning of jet fuel and furniture. Figure

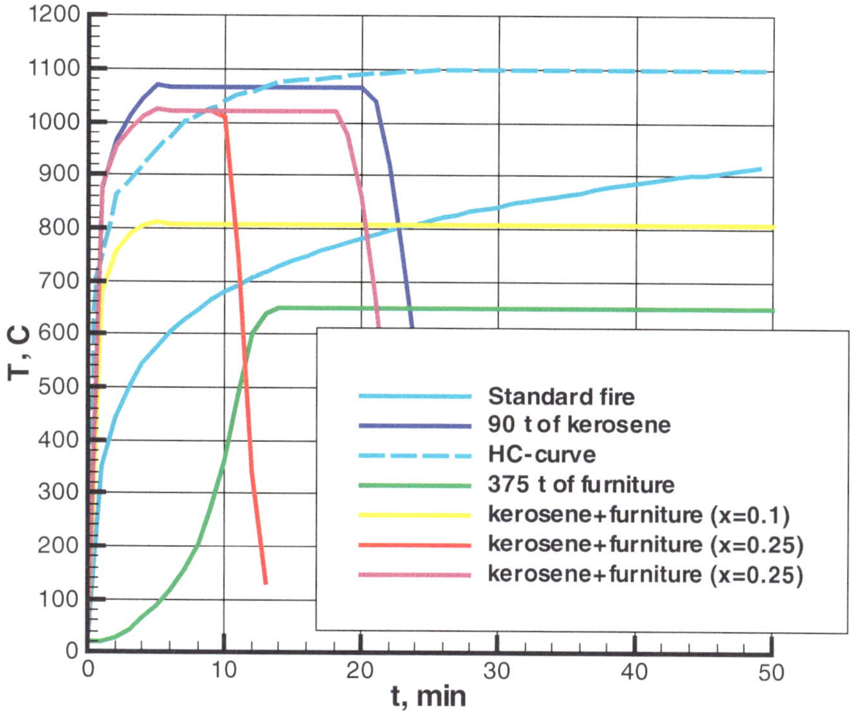

FIGURE 5-3 Calculated dynamics of average volume temperature in a structure with various amounts of jet fuel.

5-3 shows the calculated dynamics of the average volume temperature in the structure given various amounts of jet fuel assumed in the fuel load.

Based on the results of calculations of the joint combustion of spilled jet fuel and furniture, a quantitative estimate was made of the amount of fuel involved in the fire at the World Trade Center. This assessment agrees with the data from American researchers on the quantity of fuel onboard the planes just before impact.

The development of a fire may be described in more detail with the help of zone models, which are based on the premise of the formation of two layers in a burning structure: (1) the upper level of combustion products (smoke-filled zone) and (2) the lower level of undisturbed air (free zone). Thus, the status of the gaseous environment in zone models is evaluated through the use of average thermodynamic parameters from not one but several zones, and the zone boundaries are generally considered movable.

Zone models became widespread in simulating local fires in structures and systems of structures with relatively simple configurations and having comparable linear sizes.[13] However, creating zone models requires making a large number of simplifications and omissions based on a priori suppositions on flow structure. Such a method is inapplicable in cases where information on this structure that might be obtained experimentally is lacking, so consequently, there are no grounds for zone modeling. Furthermore, more detailed information is often required on the fire than just average parameter values for each layer (zone).

Field (computational fluid dynamics) models are more powerful and universal tools than zone models, inasmuch as they are based on a completely different principle. Several computer programs are currently available for field modeling, and they are fairly accurate in describing the rate, temperature, and concentration fields at the initial stage of a fire.[14]

Therefore, the field model is the best means of approaching fire modeling in complex and unique structures, for example, in transport tunnels. Figure 5-4 presents optical density fields for smoke in a central vertical section of the Lefortovo Tunnel, which is shallowly situated in Moscow's third transport ring. A study was carried out using a three-dimensional field model to predict the distribution of fire hazard factors in the tunnel both with and without antismoke ventilation.[15] A traffic accident involving a truck and several passenger vehicles in this 18.2 × 5.2-meter tunnel served as the emergency situation for the purposes of the model. In this scenario, maximum theoretical heat exchange intensity of 100 megawatts was reached 15 minutes after the start of the fire. It was supposed that the fire would break out at the center of the tunnel; therefore, given the symmetry, one-quarter of the actual tunnel volume was modeled. Calculations in the model covered 750 meters of the tunnel's length.

The temperature fields in the horizontal section at the height of 1.7 meters are presented in Figure 5-5. It is clear that despite the smoke filling the evacuation paths, the temperature in the working zone up to the 240-second mark does not exceed the critical level of 343 kelvins. Smoke with a temperature of 343 kelvins reaches the height of 1.7 meters at the 300-second mark (Figure 5-5e). At this moment, the distance from the center of the fire at which the evacuation path is blocked because of increased temperature is 90 meters.

Work on simulating fires at various types of facilities holds a significant place in prediction efforts at VNIIPO. Facilities involved in extracting, processing, and storing flammable and highly flammable liquids face a high risk of fire. In this regard, the institute has developed a software package to calculate fire and explosion hazard factors at such facilities. This software is intended for quantitative calculation of hazard factors and their consequences; visualization of calculation results in map format; and electronic communication of the results in the form of graphs, maps, and tables. Figure 5-6 presents a sample screenshot from this program.

In addition to its work on predicting the development of fires and emergency situations, the VNIIPO Situation Center is also developing numerical simulations

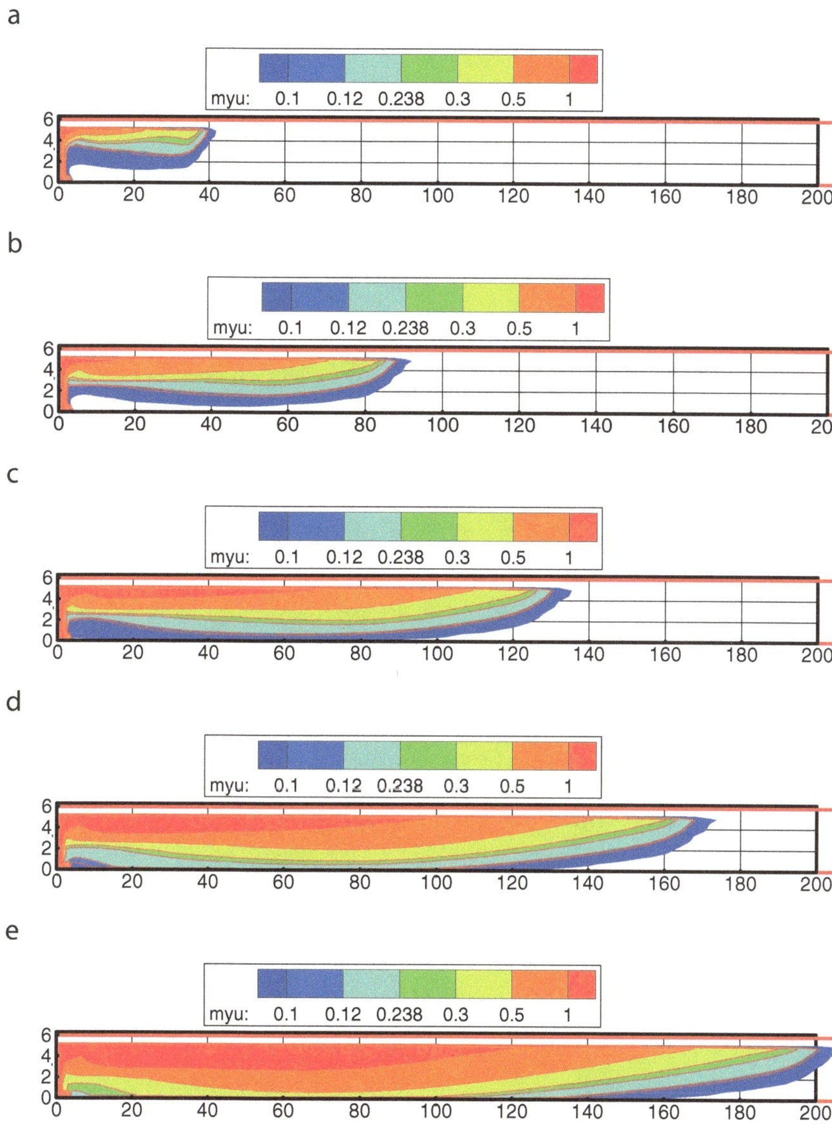

FIGURE 5-4 Optical density fields for smoke in the central vertical section of the Lefortovo Tunnel at 60 (a), 120 (b), 180 (c), 240 (d), and 300 (e) seconds after combustion.

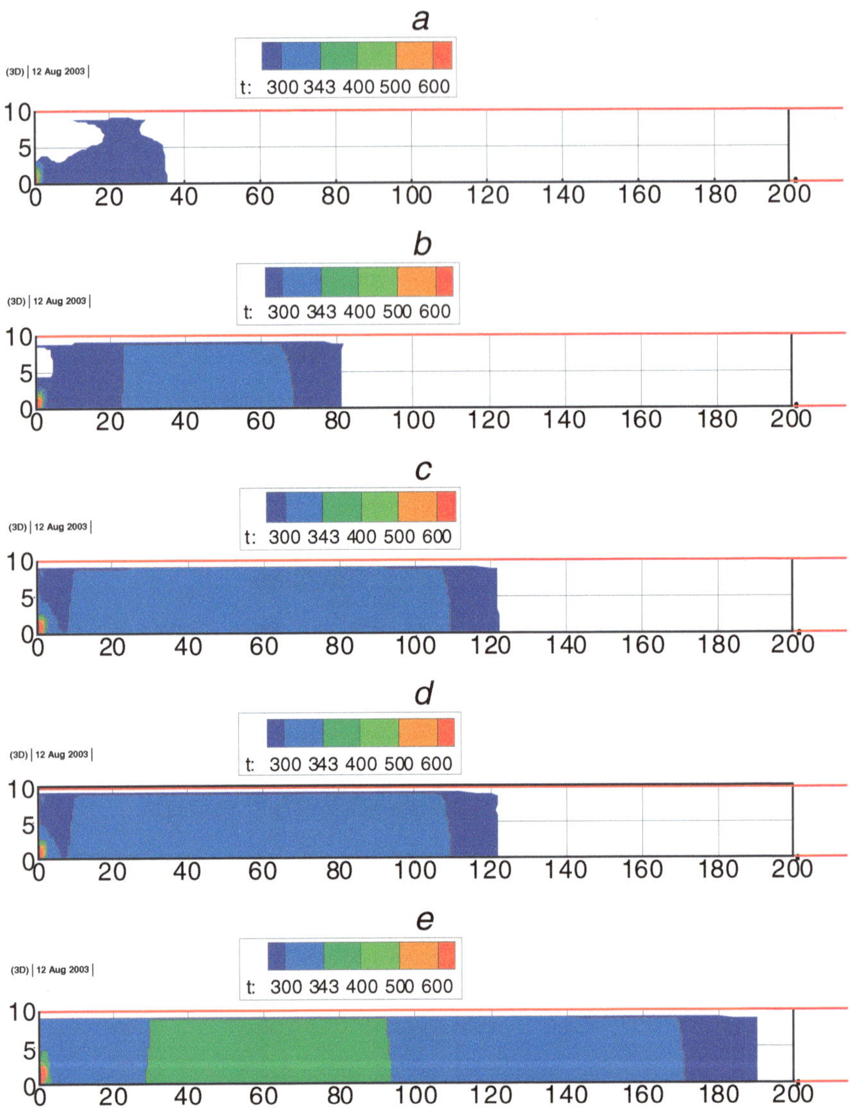

FIGURE 5-5 Temperature fields (in degrees kelvin) in a horizontal section at a height of 1.7 meters at 60 (a), 120 (b), 180 (c), 240 (d), and 300 (e) seconds after combustion.

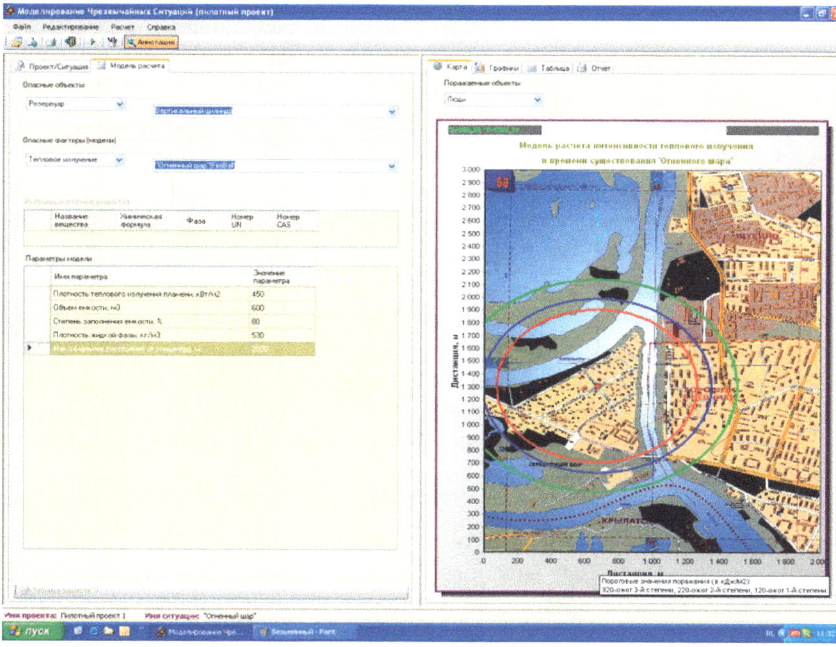

FIGURE 5-6 Sample screenshot from program to calculate fire and explosion hazard factors.

and software for process simulation of firefighting and emergency response. For example, the institute has developed a software package for calculating the personnel and resources needed to extinguish fires involving oil, petroleum products, chemicals, and stable gas condensate in storage tanks, during pour-offs to storage ponds or transfers to railway tankers, and at technical pumping stations. The program takes into account the volumes and structures of the combustion sites, the properties of the flammable liquids, tactical and technical characteristics of the foam and water delivery equipments used in extinguishing fires involving oil and petroleum products, and the characteristics of stationary and mobile firefighting equipment.

CONCLUSION

Despite existing developments in the numerical simulation of fires and emergency situations, serious issues remain to be resolved with the use of mathematical models in the work of the VNIIPO Situation Center. Based on an analysis of possible fire and emergency scenarios, a list of models in need of further refinement should be drawn up, and the need for creating new models should also be evaluated. After the models are selected, a series of studies needs to be carried out to verify them. A significant volume of work is also needed to adapt mathematical models for use in the Situation Center and to create algorithms and software packages.

Making calculations to forecast fires and emergency situations is impossible without reliable data inputs on facilities. Data collection efforts must be organized and carried out, the information must be processed, and modern technologies and geographic information systems must be used to create a database on facilities at risk of fire and explosion. In using mathematical models of fires and emergency situations based on nonlinear, nonstationary, three-dimensional model systems (for example, field models of fires), it should be taken into account that numerical solution of such systems requires tens of hours of computer time even using high-output processing technologies.

Introducing new modern technologies for numerical simulation of emergency situations requires the following:

- Improving the reliability of predictions to prevent and eliminate the consequences of emergency situations
- Organizing comprehensive monitoring and information-processing efforts regarding the status of facilities, the environment, and natural and technogenic phenomena that cause emergency situations
- Developing mathematical models of the development of fires and emergency situations
- Optimizing and facilitating timely correction of action plans and measures for preventing emergency situations as well as eliminating their consequences
- Providing a modern level of technical capabilities to support the work of operations personnel, including network communications technologies and means of collecting, analyzing, and presenting information on emergency situations

NOTES

1. *Technogenic* is used to refer to phenomena arising as a result of the development or deployment of technology.
2. Vorobyov, Yu. L. 2005. Safety in Daily Activities (Aspects of State Policy). Moscow: Business Express, 376 pp.
3. Faleev, M. I. 2002. Computer technologies in creating an information space for dealing with disasters and catastrophes. iBUSINESS 6:19-21.
4. Concept for the Creation of the National Crisis Management Center. 2005. Moscow: Ministry of the Russian Federation for Civil Defense, Emergencies, and Elimination of Consequences of Natural Disasters, 35 pp.
5. National Crisis Management Center.

6. Shatrov, V. F., and A. Yu. Silantyev. 2003. Situational centers: Information support for high-level management decisions. Pp. 8-17 in Systems Problems of Quality, Mathematical Modeling, and Information and Electronic Technologies—Part II: Imitative Modeling and Conflictology. Materials from an International Conference and the Russian Scientific School. Moscow: Radio and Communications.

7. Filippovich, A. Yu. 2003. Integration of Situational, Imitative, and Expert Modeling. Moscow: Radio and Communications, 310 pp.

8. Filippovich. Integration.

9. Romanov, V. V., and D. D. Shulga. 2003. Conceptual description of conflicting interactions. Strategic Stability 2:16-21.

10. Kopylov, N. P., A. M. Ryzhov, and I. R. Khasanov. 2000. Major fires and their modeling. Pp. 170-187 in Modeling fires and explosions, N. N. Brushlinsky and A. Ya. Korochenko, eds. Moscow: Pozhnauka [Fire Science].

Kopylov, N. P., and I. R. Khasanov. 2001. Predicting the fire situation at sites under demolition. Pp. 101-102 in Extreme Situations: Prevention and Elimination. Collected Materials from a Scientific-Practical Conference. Minsk: Belarus State University.

11. Ryzhov, A. M, I. R. Khasanov, A. V. Karpov, et al. 2003. Application of a Field Method for Mathematical Modeling of Fires in Structures: Methodological Recommendations. Moscow: VNIIPO, 35 pp.

12. Astapenko, V. M., Yu. A. Koshmarov, I. S. Molchadsky, and A. N. Shevlyakov. 1988. Thermodynamics of Structure Fires. Moscow: Stroiizdat [Construction Publishers], 448 pp; Molchadsky, I. S. 2005. Fire in a Structure. Moscow: VNIIPO, 456 pp.

13. See, for example, Cooper, L. Y., J. A. Rockett, H. E. Mitler, and D. W. Stroup. 1989. A program for the development of a benchmark compartment fire model computer code. Fire Technology 25(4):116-127.

Takeda, H. 1988. Transient model of early stages in compartment fires. Pp. 21-34 in Mathematical Modeling of Fires, J. R. Mehaffey, ed. Philadelphia: ASTM; Merkushkina, T. G., and V. V. Romanov. 1981. Use of mathematical modeling in studying fire hazard factors. Pp. 34-43 in Safety of People in Fires. Moscow: VNIIPO.

14. Ryzhov et al. Mathematical Modeling of Fires.

Ryzhov, A. M. 2000. Field models of fires. Pp. 25-88 in Modeling Fires and Explosions, N. N. Brushlinsky and A. Ya. Korolchenko, eds. Moscow: Pozhnauka.

Yang, K. T., J. R. Lloyd, A. M. Kanury, and K. Satoh. 1984. Modeling of turbulent buoyant flows in aircraft cabins. Combustion Science and Technology 39:107-118.

Raycraft, J., M. D. Kelleher, H. Q. Yang, and K. T. Yang. 1990. Fire spread in a three-dimensional pressure vessel with radiation exchange and wall heat losses. Mathematical and Computer Modeling 14:795-800.

Cox, G. 1995. Combustion Fundamentals of Fire. London: Academic Press, 476 pp.

Welch, S., and P. Rubini. 1996. SOFIE—Simulations of Fires in Enclosures: User Guide. Bedford: Cranfield University, 127 pp.

15. Ryzhov et al. Mathematical Modeling of Fires.

6

Organizational Measures and Decision Support Systems for Preventing and Responding to Terrorist Acts at Potentially Hazardous Facilities, on Transportation Systems, and in Locations Where Large Numbers of People Congregate*

A. Yu. Kudrin, Director, All-Russian Scientific Research Institute for Civil Defense and Emergency Situations (ICDES) (Federal Center); A. I. Zaporozhets, Deputy Director for Research, ICDES; and S. A. Kachanov, Deputy Director, ICDES

Terrorism is a complex, multifaceted phenomenon that is social in nature and, in some instances, has a political aim. Terrorists attempt to exert political pressure on government leaders, attract world public attention to certain problems, demand the liberation of arrested supporters of extremist groups and the end of persecution of terrorist organizations and their leaders by law enforcement agencies, advance economic demands, and so forth.

As a rule, terrorists commit individual acts of an intentionally provocative nature, which may include threats of murder or the assassination of state and political figures; the seizure of hostages or potentially hazardous facilities; bombings; or the release of poisons, radioactive substances, or biologically active agents. This will lead to deaths among members of the public who happen to be at the site of the attack and will harm the economy and the prestige of the state.

Terrorist acts at potentially hazardous facilities—enterprises working with chemicals, radioactive materials, or explosives; hydrotechnical structures; unique tall buildings; subways, surface rail, and air transport facilities; and places where large numbers of people congregate, such as concert halls, stadiums, apartment

*Translated from the Russian by Kelly Robbins.

buildings, and so forth (hereafter referred to as facilities)—present a great danger to personnel and the public and cause substantial economic damage. Terrorist acts at enterprises could be carried out by striking (destroying) a tank or pipeline holding catastrophically hazardous chemicals, a nuclear reactor, or a storage vessel containing highly flammable liquid. An explosion at a chemical-hazard facility could cause destruction over an area of up to 30 square kilometers, with the number of injured victims possibly reaching 60,000 and up to 5,000 fatalities. Destruction of an atomic reactor could contaminate up to 1,200 square kilometers, with the number of casualties in this situation possibly reaching 10,000.

If a terrorist strike against a hydrotechnical structure were to occur, we might expect cities or towns to be flooded and buildings to be destroyed by the resulting surge of water. The land area submerged could reach about 1,000 square kilometers, with the number of victims possibly reaching 120,000.

In places where large numbers of people gather, terrorists could use explosives, dangerous chemicals (including poisons), radioactive substances, and biologically active agents.

The suddenness of a terrorist act, the rapid spread of the impact factors, the deaths of many people, the ensuing panic, and people's sense of being unprotected create a powerful psychological blow to society. Therefore, prompt response to a terrorist threat or act is an important factor in preserving the lives and health of people subjected to such attacks. Many organizations of various types have been involved in studies on preventing and eliminating the consequences of terrorist acts. The Russian Ministry for Civil Defense Affairs, Emergencies, and Elimination of Consequences of Natural Disasters (EMERCOM) is responsible for emergency rescue and other urgent efforts involved in eliminating the consequences of terrorist acts.

A unified state system for emergency situation prevention and response has been created and is operating in Russia. Within the boundaries of specific jurisdictions, administrative agencies specially empowered to handle issues related to protecting the public and area from emergency situations, depending on the circumstances and scope of the predicted or actual situation, establish one of the following operating regimes for the subsystems of the unified state system for emergency management:

- Standard daily operating regime: during normal production activities in the absence of any predictions of possible terrorist acts
- Increased readiness regime: when possible terrorist acts are predicted at a facility
- Emergency regime: when a terrorist act has been committed at a facility

Basic measures involved in the standard daily operating regime are as follows:

- Situational observation and monitoring at facilities and adjacent areas by facility staff and law enforcement personnel
- Organization and implementation of training for local government and law enforcement personnel, facility staff, the public, and emergency rescue personnel in means of protection and appropriate actions to be taken at a facility, on a transport system, and in open spaces if a terrorist act is committed
- Planning, organization, and implementation of training exercises on emergency warnings, protection of people against the effects of impact factors, and reduction of losses and damage from a terrorist act
- Participation in the development and implementation of organizational and engineering-technical measures to ensure more stable operations of facilities and transport systems in an emergency resulting from a terrorist act
- Creation and augmentation of stores of emergency supplies and monitoring of the usability of individual protective gear; medical supplies for individual protection; and equipment needed for communications, public notification, and chemical, radiation, and biological surveillance and monitoring
- Organization of matters regarding the interactions of special emergency response subunits with EMERCOM, the Ministry of Internal Affairs, the Ministry of Communications, the Federal Security Service, the Ministry of Healthcare and Social Development, the Ministry of Defense, and other Russian ministries and departments
- Cooperation with local government agencies and officials specially authorized to deal with combating terrorist acts in order to select sites for decontamination stations for equipment and clothing, sanitary washing stations, and accumulators (observation stations) for eliminating the consequences of terrorist acts at facilities that involve the use of dangerous chemicals (poisons), radioactive substances, and biologically active agents
- Training of personnel from EMERCOM specialized subunits and facility staff on actions to eliminate the consequences of terrorist acts, including victim assistance and use of technical means for containment, special processing techniques using equipment, and sanitary processing of individuals
- Other matters aimed at preventing losses and reducing damage from a terrorist act in accordance with the specific characteristics of operations at each particular facility

Basic measures involved in the increased readiness regime are as follows:

- Assumption by the appropriate emergency commission of direct operational command of the emergency management system subunit functioning at the site of the terrorist act; formation of operations groups at the local level to ascertain the situation at the site of the terrorist act; and provision of effective assistance to facility staff and law enforcement personnel in dealing with the emergency

- Communication of the threat (prediction) of an emergency situation to the appropriate Russian Federation ministry, department, or organization with jurisdiction
- Notification of facility staff and law enforcement personnel at the facility about the terrorist act
- Testing of operational communications and clarification of interactions between the appropriate emergency commission and the EMERCOM crisis management center
- Augmentation of security and dispatch services at the facility
- Increased observation of the situation at the facility and in adjacent areas
- Distribution of individual respiratory protective gear to be kept at the ready by facility staff and law enforcement personnel
- Preparation of portable and mobile devices for chemical, radiation, and biological surveillance and monitoring for use if needed
- Placement of EMERCOM personnel and resources at the appropriate level of readiness; clarification of plans for their actions

Basic measures involved in the emergency regime are as follows:

- Clarification of the situation in the zone where the terrorist act occurred
- Notification of officials specially authorized for involvement in combating terrorist acts, facility staff, and law enforcement personnel that a terrorist act has been committed
- Conduct of overall survey (chemical, radiation, or biological, as needed) and monitoring efforts to establish whether explosives, hazardous chemicals (poisons), or biologically active agents were used; establishment of perimeter of zone impacted by explosives, affected by chemicals (poisons) or biological agents, or contaminated by radioactive substances
- Evacuation of the public from the danger zone
- Distribution of individual protective gear to victims (if necessary)
- Provision of initial medical and paramedical assistance to victims
- Execution of measures to eliminate the source of the emergency
- Execution of measures to decontaminate the area; specially process individual protective gear, uniforms, and equipment; and provide sanitary processing for personnel involved in containing and eliminating the consequences of terrorist acts in which dangerous chemicals (poisons), radioactive substances, or biologically active agents were used

Based on an analysis of likely threats that could lead to emergency situations, individualized security systems must be developed for industrial enterprises, unique tall buildings, facilities where large numbers of people gather, and subway stations. Such systems should take into account natural, technogenic, biological, social, and terrorist factors that could cause emergency situations. Such security systems include both organizational and technical measures. Organizational measures provide plans for the actions of personnel, residents, and facility visitors both in regular day-to-day activities and during emergencies, threats of terrorist acts, and such acts themselves. They are laid out in the appropriate regulatory technical documents. Plans for rescuing and evacuating people and eliminating the source of the emergency should be developed in a timely fashion through training exercises and courses.

Technical measures are developed with the aim of supporting the normal functioning of a facility under its regular daily operating regime, during threats, and during actual emergencies. These measures are implemented using engineering and technical means: design and structural elements; barriers; blocking devices and mechanisms; security, fire alarm, and warning systems; systems for monitoring and management of facility security and critical operations; loudspeakers and other means of notification; video observation systems; means and systems for facility access control and management; environmental monitoring equipment; and so forth.

All facilities must be prepared for an emergency. To achieve this, measures are taken to improve the facility's level of protection. The list of measures could be augmented and revised depending on the facility's function.

Critically important points are identified in the design stage. When the facility is in operation, access to these points is limited and they are constantly monitored. Such points include structural elements that, if destroyed, would lead to destruction of the entire facility, as well as technological systems and equipment that, if affected by an accident, could lead to an emergency situation. Physical protection boundaries are organized and equipped with monitoring rooms, alarm systems, controlled access points, and inspection points for vehicles and individuals. The necessary badges or access cards are distributed to staff. An accounting is made of residents and visitors. Regulatory technical documents on actions to be taken in an emergency and systems for communications with supervisory agencies and fire and rescue personnel are developed and forwarded to those responsible for implementation. Special attention is devoted to seeking and detecting unauthorized persons and objects, finding them in a timely manner, and handing them over to law enforcement personnel or specialists. Personal cars and trucks with compressed gas-powered engines are prohibited on the grounds of the facility. The number of persons involved in facility access control and monitoring is increased.

As for preventing and eliminating the consequences of emergency situations at facilities, much attention is currently being focused on automated systems used there. The All-Russian Scientific Research Institute for Civil Defense and Emergency Situations (ICDES) has developed an original technology and the necessary regulatory and methodological base for creating automated interconnected security and utility systems and structured systems for monitoring and managing engineering systems at buildings and structures. The technology that has been developed makes it possible to prevent or significantly reduce the consequences of emergency situations caused by critical utility failures; the sudden collapse of structural elements of buildings and other structures; fires; explosions; increased levels of hazardous chemicals, radiation, or biologically active substances; or terrorist acts.

The programmatic and technical solutions that have been developed make it possible to

- support the operations of all security and critical utility systems according to previously determined algorithms in emergency situations, including those caused by terrorist acts;
- facilitate the uninterrupted, remote, real-time, automated processing of information on the status of critical utility and security systems and engineering-technical elements at facilities and automatically transmit the necessary data on the parameters of the emergency in an established format to the necessary response service; and
- facilitate long-distance management of critical utility and security systems at facilities from a remote control center in the city in emergency situations, including those caused by terrorist acts.

Preliminary calculations indicate that the creation of this system would reduce the number of emergency situations in buildings and structures by at least 20 percent and would cut materials losses by more than 15 percent.

To prevent the sudden collapse of buildings and structures, the institute has developed a technology for remote monitoring of engineered technical elements. Two parameters are monitored: (1) individual fluctuation periods (frequencies) of barrier elements and (2) displacements (vertical, horizontal, and twisting). The data obtained are then automatically processed mathematically, resulting in output on the condition of the structural elements of the facility (normal, increased attention, or alarm). The results of the monitoring are automatically transmitted to the necessary response services.

With the aim of preparing scientifically grounded recommendations on actions to be taken in emergency situations, including those caused by terrorist acts, the Center for Decision Support in Emergency Situations has been created at ICDES. The center has the necessary software allowing it to determine automatically the scope of a given emergency and to prepare the necessary recommendations on rescue and other urgent efforts to save people and minimize material damages. The software was created on the basis of existing packages and well-proven methodologies. Measures to prevent and respond to emergencies are being prepared not only on the basis of theoretical tasks but also using accumulated knowledge bases such as previous experience in carrying out such efforts, new methods and technologies for eliminating the consequences of various types of emergencies, data on new emergency rescue tools and their availability, and so forth.

Specialists from other ministries and departments could be involved in preparing recommendations, including by means of videoconferencing.

The center's preparation of timely, scientifically grounded recommendations on preventing and responding to various types of emergency situations makes it possible to reduce substantially the likelihood of a disaster at facilities and to save significantly more people and reduce the costs of rescue efforts if such situations do occur, including those caused by terrorist acts.

7

Characteristics of Technological Terrorism Scenarios and Impact Factors*

Nikolai A. Makhutov, Vitaly P. Petrov, and Dmitry O. Reznikov,
Russian Academy of Sciences Institute of Machine Sciences

INTRODUCTION

Technological terrorism is defined as actions directed against infrastructure elements critically important for national security or committed with the use of especially hazardous technologies, technical means, and materials. In considering technological terrorism scenarios, the primary impact factors of such terrorist acts initiate secondary catastrophic processes with a significantly higher (tens and hundreds of times) level of secondary impact factors that affect the targets of the attack, their personnel, the public, and the environment.

The scope and intensity of the impact factors of terrorist actions against a given system define the level of the terrorist threat to that system.

The scenario for a terrorist attack entails a means of exerting the initiating effect on the system that is based on the use of appropriate technical devices, technologies, and materials and is characterized by the terrorists' deliberate selection of the place and time of the attack.

The following characteristics must be taken into account in analyzing technological terrorism scenarios and impact factors.[1]

*Translated from the Russian by Kelly Robbins.

High level of dynamism: Terrorist attack scenarios and impact factors are more dynamic in nature than scenarios and impact factors for natural and technogenic[2] disasters to which the system is subject. Of course, emergency management and evacuation capabilities are relevant to both. A change in the spectrum and intensity of possible terrorism-related extreme effects on the system is significantly more powerful than a natural or technogenic threat. This is due to the terrorists' capacity for constantly expanding their arsenal of mechanisms for initiating emergency situations using modern means of attack, reacting to changes in protection systems, and drawing lessons from mistakes made during previous attacks on the system or others like it.

High level of uncertainty: In modeling terrorist scenarios and impact factors, we encounter a higher level of uncertainty. In addition to the undefined factors inherent in threats of a natural or technogenic nature, terrorist threats entail new factors of uncertainty resulting from the complexity of evaluating terrorists' value system and behavioral logic as well as their organizational-technical potential and the resources at their disposal.

Capability of terrorists to choose attack scenarios deliberately: This refers to terrorists' deliberate selection of attack scenarios (places, times, and types of actions), taking into account system vulnerability parameters and the damages expected if an attack is successfully carried out. That is, terrorists are capable of analyzing the vulnerability matrix and damage structure for various types of actions against a system and selecting the attack scenario that maximizes the harm to society (taking secondary and cascade damages into account). Here, in addition to probability analysis, it is also necessary to apply the tools of game theory, which makes it possible to take the intentional actions of terrorists into account.

Characteristics of the perception of the terrorist threat: A significant part of the population is inclined to fear terrorist attacks to a greater degree than equivalent natural and technogenic phenomena as described in the equation $R = H^n \cdot V \cdot U$ where $n < 1$, the indicator for the degree characterizing the subjective perception of the consequences of terrorist acts.

Complex nature of the terrorist threat: The presence of a terrorist organization in a region may give rise to the possibility of a broad spectrum of attack scenarios, including the time, place, and character of the attack. Thus, to counter terrorist threats and terrorist mechanisms for initiating emergency situations to an even greater degree than for natural and technogenic risks, a complex systems approach is needed for ensuring security and developing an optimal strategy for counterterrorism force and resource deployment. Inasmuch as concentrating resources on protecting one system element (or protecting a target from one type of terrorist action) could prove useless because, after evaluating the situation, the terrorists could either redirect the attack against another element of the target or could switch to a different type of attack. In this case, counterterrorism efforts will not lead to reducing risk and increasing the target's level of protection.

In addressing traditional tasks of ensuring security against natural and technogenic disasters, the prevailing types of impact factors could be highlighted for the system being studied, such as threats from seismic activity, flooding, chemical contamination, and so forth. In protecting the system from these impact factors, it is possible to achieve the desired result. However, in protecting a given system from manifestations of terrorism, the spectrum of potential threats is significantly wider. Here, terrorists are capable of analyzing the level of protection of the system for various types of impact factors, identifying impact factors against which the target is least protected, and concentrating their efforts on carrying out an attack that will bring these very factors to bear.

Furthermore, there are types of terrorist actions with no analogues in the structure of impact factors typical of natural and technogenic disasters: for example, cyberterrorism or electromagnetic actions aimed at knocking control systems out of commission.

Global nature of terrorist threats: As a rule, the geographic distribution of sources of natural and technogenic threats is limited to regions where hazardous facilities are located or zones subject to natural hazards (river valleys for floods, seismic fault zones for earthquakes, tsunamis, and so forth). On the contrary, terrorist threats, especially those coming from international terrorist networks, are characterized by significantly more widespread distribution of the locations where a possible attack might occur.

Presence of aftereffects in the flow of terrorist actions: In contrast to natural and technogenic disasters, which may often be viewed as chains of Poisson events, after a major terrorist act the condition of the system defined as "terrorist organization—protected object—protection system" is substantially changed. On the one hand, the terrorist organization achieves its goals to one or another degree and expends a significant part of its resources, while, on the other hand, law enforcement agencies intensify the protection regime. Therefore, after a major terrorist act the situation fundamentally changes and the likelihood of a subsequent attack is significantly altered as well (generally, it is reduced). Therefore, the sequence of terrorist attacks could be described with the help of a Markov chain model. For the purpose of this model, the activities of antiterrorist forces aimed at countering the terrorist threat are understood as under control. The Markov process model makes it possible to describe the dynamics of cycles of terrorist activity.

Terrorists' capacity for self-learning: Because terrorists are capable of analyzing the results of previous attacks and drawing conclusions from them, their experience in "successful" and "unsuccessful" attacks can have a noticeable effect on the selection of a scenario for the next attack. (Attack scenarios that have proven their effectiveness in the past have a great likelihood of being repeated by terrorists in the future, while scenarios that ended unsuccessfully will most likely be less attractive to terrorists and consequently are less likely to be repeated.) Therefore, in assessing the chances that various attack scenarios will be realized, statistical self-learning models are more effective than traditional frequency methods.

Presence of two-way linkages between the terrorist threat and system vulnerability: One differentiating feature of a terrorist threat to a given system is the presence of two-way linkages between that threat and (1) vulnerability of the system to that threat and (2) the magnitude of expected damages if the threat is successfully realized. This characteristic of terrorist mechanisms must be examined in more detail, inasmuch as it opens up additional possibilities for reducing terrorism risks.

The formula for assessing the risk of a traditional emergency situation initiated by a natural or technogenic disaster could be presented in simplified form as follows:

$$\boldsymbol{R}_c = \boldsymbol{P}_{IV} \times \boldsymbol{P}_{(NU/IV)} \times \boldsymbol{U}_{(damage/IV\ \&\ NU)}$$

Here $\boldsymbol{P}_{IV}$ is the threat to the system, expressed as the probability of an extreme initiating action (the failure of a particular element, exceeding of allowable level for a hazard factor, extreme natural phenomenon, and so forth).

$\boldsymbol{P}_{(NU/IV)}$ is the vulnerability of the system to the given initiating action, expressed as the conditional probability that damage will be inflicted if the initiating action occurs.

$\boldsymbol{U}_{(damage/IV\ \&\ NU)}$ is the damage inflicted on the system if the initiating action occurs and causes damage.

Thus, for traditional natural and technogenic disasters, vulnerability is determined by a specific threat, but the consequences depend on both the type of threat and the vulnerability of the system to that type of threat. Here it should be noted that in this model there are no two-way linkages, such as the dependence of the threat on vulnerability (inasmuch as the probability of a spontaneously initiated action has no relation to system vulnerability to that action) or dependence of the threat on the consequences (for the same reason).

Therefore, the system of linkages among the risk factors for the given system in an emergency of a natural or technogenic nature is as presented in Figure 7-1A.

If the initiating action is a terrorist attack, the interactions among the various factors included in the risk assessment equation are more complex. Similar to the expression above, terrorism risk is presented as follows:

$$\boldsymbol{R}_T = \boldsymbol{P}_A \times \boldsymbol{P}_{(NU/A)} \times \boldsymbol{U}_{(damage/A\ \&\ NU)}$$

$\boldsymbol{P}_A$ is the terrorist threat to the given system, expressed as the probability that a terrorist attack of a particular type will be carried out.

$\boldsymbol{P}_{(NU/A)}$ is the vulnerability of the system to a terrorist attack of the given

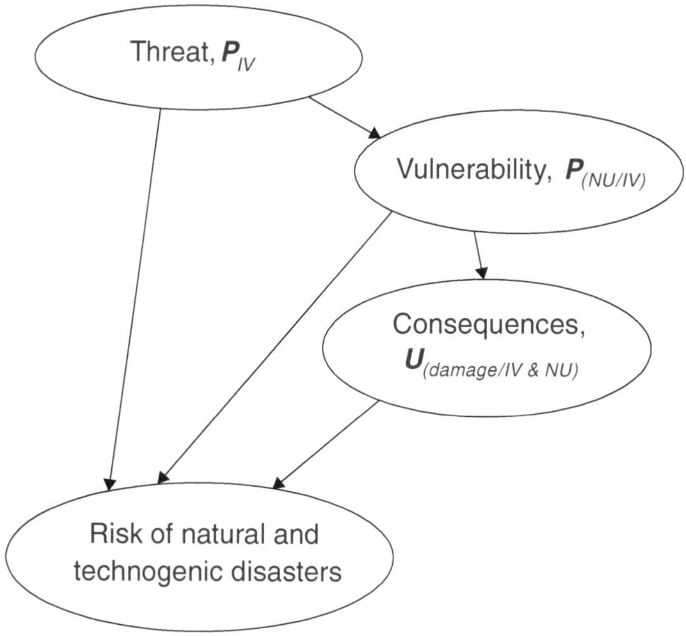

FIGURE 7-1A System of linkages among risk factors for emergency situations of a natural or technogenic nature.

type, expressed as the conditional probability that damage will be inflicted if the attack is carried out.

$U_{(damage/A \& NU)}$ is the damage inflicted on the system if the terrorist attack is carried out and causes damage.

If a terrorist action occurs, the presence of powerful two-way linkages among the risk factors should be noted (see Figure 7-1B).[3] In particular, reducing the vulnerability of a given system makes it possible to reduce substantially the level of the terrorist threat it faces.

MAIN TYPES OF SCENARIOS AND IMPACT FACTORS FOR TERRORIST ACTIONS

Based on an analysis of the growing number and expanding spectrum of terrorist actions, we may conclude that scientific-technical progress presents terrorists with new opportunities for carrying out various types of terrorist acts. Successes in the development of advanced technologies and means of communication, high rates of urbanization, and the concentration of potentially hazardous

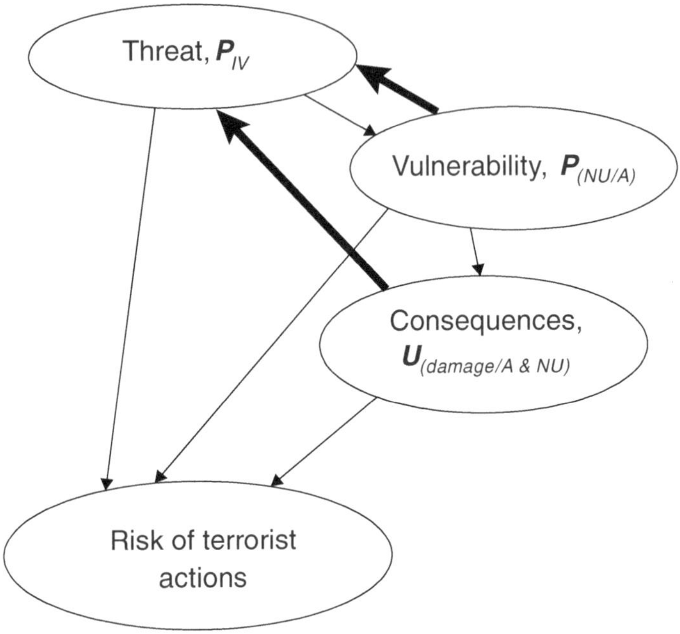

FIGURE 7-1B System of linkages among risk factors for emergency situations of a terrorist nature.

production facilities create favorable conditions for the appearance of new types of technological terrorism with especially dangerous consequences for the public and government institutions.

On the other hand, scientific-technical progress also makes it possible to protect the public and objects in the technosphere from terrorist actions. It is technical means of protection that provide the possibility of preventing terrorist acts and minimizing their consequences; that is, they make it possible to protect critically important targets, personnel, the public, and the environment.

The following section will cover the main types of scenarios for technological terrorism.

Electromagnetic Terrorism Scenarios

Modern critically important facilities (ground- and space-based communications systems, telecommunications systems, computer networks, power plants, transport control systems, nuclear industry facilities, and so forth) are vulnerable to the impact of powerful electromagnetic irradiation and penetrating high-voltage electrical pulses in electricity supply and grounding networks. This circumstance has led in recent years to the appearance of a real danger that scenarios for terrorist attacks based on the application of electromagnetic effects may in fact be realized.[4]

Electromagnetic terrorism scenarios entail the intentional use of electromagnetic effects against electrical and electronic systems to disrupt their normal operations. The foundations for the rise of the threat of electromagnetic terrorism were laid, on the one hand, by the sharp reduction in signal levels in electronic systems and, on the other hand, the sharp growth in achievements in creating pulse flow generators and, on their basis, electromagnetic wave emitters. The widespread introduction of electronic systems in all spheres of societal activity and the accessibility of devices used to create disruptions have given rise to the very real threat that electromagnetic terrorism scenarios may be implemented.

Electromagnetic terrorism scenarios may be divided into four main groups:

1. Injection of electrical field pulses into the electricity supply networks serving electronic devices and information systems
2. Use of super-broadband emissions to affect electronic devices and control systems
3. Creation of electromagnetic clouds to damage electric transmission lines
4. Use of electromagnetic emissions to detonate mines or other explosive devices placed for the purpose of sabotage

Cyberterrorism Scenarios

The development of computer networks and information systems based on packet commutation technology has created a new communications and information environment that is vulnerable to terrorist acts.[5] Attacks by computer terrorists could be aimed at specific elements of the information infrastructure itself, possibly by means of computer networks, or at other targets present in one way or another in this environment. The network infrastructure as such could be of enormous value to terrorists, inasmuch as it provides a cheap and effective means of interaction and communication and serves as a source from which information may be obtained.

Thus, in addition to the multitude of positive aspects, the development of cyberspace significantly expands terrorists' arsenal of tools and capabilities. The possibilities offered by global network technologies allow terrorists to work in practically any country against targets located in any other country.

In modern industrially developed society, information technologies may be viewed by terrorists as both a target of attack and a means of attack.

Not only telecommunications and information networks but also all other components of any vitally important (critical) infrastructure whose successful functioning depends on computer control, data processing, and digital communications could become targets for cyberattacks.

The following are highlighted as main scenarios for computer terrorism: destroying network infrastructure on a corporate, national, or transnational scale by knocking their control systems or individual subsystems out of commission; obtaining access to confidential data; changing data affecting the outcome of processes in which terrorists have an interest; or an information-related action resulting in individuals or groups behaving in accordance with terrorists' wishes.

Computer terrorism scenarios may be particularly effective if used in combination with physical actions against critically important targets. In such cases, a cyberattack is used as a factor intensifying the effect of the physical attack by countering the efforts of rapid-response services and communications and command systems, providing false output data that cause leaders and personnel to take inadequate actions, or creating panic among the public. Thus, cyberattacks increase the danger from a physical attack and exacerbate its consequences by complicating response actions and implementation of damage reduction measures.

Biological Terrorism Scenarios

The impact factors of biological terrorism can cause massive disease outbreaks and panic among people, animals, and plants.[6] These impact factors include microorganisms and some of their products (toxins), as well as certain types of insects, both plant pests and disease vectors. In means of application, bioterrorist acts differ from other types of terrorist acts in that they can be both overt, announced, demonstrative acts as well as covert acts masked as natural outbreaks. Here it should be noted that according to current information, a significant portion of cases of bioterrorism are covert or masked in nature. Therefore, the problem of differentiating natural and artificially created disease outbreaks remains very urgent.

The effectiveness of biological terrorism scenarios is determined by the following factors:

- The world is currently witnessing the rapid development of the biological sciences, biotechnology, medicine, and pharmacology. Increasing numbers of people are employed in these fields and have the necessary knowledge and qualifications to develop and manufacture bioweapons. There is a growing number of laboratories and biological and pharmaceutical plants that have the necessary conditions for producing biological weapons.
- Manufacturing biological weapons is relatively simple and inexpensive. With the appropriate pathogenic virus or microorganism strain, a pathogen can be produced in rather large quantities without particular problems in any laboratory

with the capacity to support work under sterile conditions. Such conditions are relatively easy to create even at home.

- The problem of the pathogenic bacterial or viral strain is that although it is one of the most complex problems bioterrorists face, it is also solvable. Bioterrorists can obtain the pathogen illegally through a laboratory or production facility where these microorganisms or viruses are studied or where related vaccines or diagnostic test kits are produced. Bioterrorists can then pass the pathogenic viral or microbial strain along to another terrorist group.
- Bioweapons are effective in very small doses. The ease of concealing bioweapons, the covert manner in which they can be used, the lack of external manifestations at the moment of release, and the relative ease with which they can be produced make it very unlikely that their use will be detected and prevented.
- Biological weapons make it possible to carry out both individual terrorist acts and massive strikes against people, animals, and plants.
- At present, there are practically no technologies for protecting against bioweapons or detecting and identifying a pathogenic microorganism or toxin before it begins to take effect. Therefore, a case of bioterrorism can be discovered only after the outbreak begins and is identified, which can take a fairly long time after large numbers of people, animals, or plants have already been infected.

Thus, the relative ease of producing biological weapons, the practical invulnerability of the perpetrators, and the possibility of damages on a huge scale make biological attack scenarios attractive to terrorists.

Chemical Terrorism Scenarios

Dangerous chemicals are found everywhere in modern industrial society and, consequently, may be accessible to terrorists.[7] The following four attack scenarios related to chemical terrorism may be highlighted:

1. Dispersal of a military chemical substance for nonmilitary purposes
2. Sabotage at a chemical plant or storage facility (including rail tank cars) where there are toxic chemicals stored in gaseous, liquid, or solid form that can react with air or water to produce toxic gases or evaporate into the atmosphere
3. Contamination of natural water sources or drinking water reservoirs with toxic substances
4. Intentional use of chemical substances to kill individual people

Terrorists may realize their intentions of acquiring chemical weapons in two ways: (1) by buying (stealing) them from existing national stockpiles or (2) by producing them at their own underground enterprises. Inasmuch as synthesizing military chemical substances requires overcoming complex technical barriers and entails great risk, it is more likely that terrorists will acquire highly toxic industrial chemicals. Although such substances are hundreds of times less lethal than paralyzing nerve gas, they nevertheless can cause significant losses if used in a closed space or in the open air under favorable atmospheric conditions.

Military chemical substances are poisonous, artificially created gases, liquids, or powders that upon entering the body through the lungs or skin cause disability or death among people and animals. Although many military chemical substances are liquids, they can be put into the form of an aerosol (fine mist of tiny droplets) and then evaporate into the atmosphere as a result of the detonation of a shell. Most chemical substances fall into one of five broad categories: (1) skin-blistering agents, (2) paralyzing nerve agents, (3) asphyxiating gases, (4) bleeding agents, and (5) disabling agents. Besides the various psychological effects that they produce, chemical weapons also differ from one another in their resistance to destruction, volatility, and evaporation rate. Unstable substances are dispersed in the air for several hours and mainly present a threat if they are inhaled, while persistent substances remain dangerous for a month if they are scattered on the soil, vegetation, or objects and, as a rule, represent a hazard if they make contact with skin.

Chemical substances with skin-blistering effects, such as yperite (mustard gas) or lewisite, are liquids that cause chemical burns.

Nerve-paralyzing substances like sarin and VX are the most powerful chemical poisons known. They disrupt the human nervous system and kill their victims within a few minutes. Given the extreme danger associated with handling or storing nerve-paralyzing agents, terrorists might attempt to develop a binary weapon that would be safer to produce, store, and transport. A binary system presumes the separate storage of two relatively nontoxic ingredients and their mixture immediately before use to create a lethal substance. Sarin, for example, could be produced in a binary system through the chemical reaction of isopropanol with methylphosphoryldifluoride (DF). However, synthesizing DF is complicated and difficult. Furthermore, terrorists would have to either mix the components manually before use, which is an extremely dangerous operation, or try to develop a remote-controlled device to handle the mixing and dispersal, which in turn would require a high degree of technical skill.

A very likely terrorist attack scenario would be a case of sabotage at an industrial enterprise that manufactures, processes, or stores highly toxic chemicals, leading to their emission or discharge with subsequent impacts on nearby populated areas. A dangerous chemical could be intentionally discharged by destroying a chemical container with the help of a conventional explosive device or by sabotaging the manufacturing process at the facility, leading to an emergency situation. Terrorists could also set off an improvised explosive device to blow a hole in a rail tank car being used to transport a dangerous chemical.

Historical experience shows that most well-known chemical terrorism attacks were committed using household and industrial chemicals. Although these substances are less toxic than military poisonous substances, their consequences may be catastrophic. Therefore, in addition to countering attack scenarios using military poisonous substances, it is recommended that significant attention be devoted to attack scenarios involving sabotage at facilities producing, using, or transporting hazardous chemicals.

Radiation Terrorism Scenarios

Scenarios for terrorist acts using radiation sources may be divided into three groups: (1) detonation of a nuclear explosive device, (2) sabotage at nuclear facilities, and (3) radiological terrorism.[8]

Detonation of a Nuclear Device

Scenarios in this group relate to a more dangerous type of terrorism from the standpoint of the scope of the consequences. Such scenarios entail the theft of a nuclear explosive device from a storage arsenal or the creation of a homemade nuclear bomb using highly enriched uranium or plutonium. Realization of these scenarios is complicated by the circumstances that the key components necessary for manufacturing nuclear weapons systems—that is, fissible materials (plutonium or highly enriched uranium)—are difficult to obtain, and the capabilities and equipment needed to produce them also have their specific characteristics. However, although nuclear weapons systems are complex technical devices, it is impossible to rule out the possibility that a well-trained terrorist organization could be capable of manufacturing a primitive nuclear device with a yield up to the tens of kilotons.

The most difficult part of manufacturing such a nuclear device is acquiring the necessary quantity (on the order of several kilograms) of highly enriched uranium or plutonium. Therefore, preventing nuclear weapons and weapons materials from falling into the hands of terrorists is a top priority.

Sabotage at Nuclear Facilities

Scenarios in this nuclear terrorism category entail setting off an explosion at a facility such as a nuclear power plant, research reactor, spent fuel reprocessing plant, radioactive waste repository, or similar site.

Numerous nuclear facilities present very attractive targets for terrorists. The potential destruction and damage that could be caused by a terrorist act at a nuclear reactor depend on the design characteristics of the given reactor and the protective measures in place, which in turn vary widely at the different types of facilities. According to data from the International Atomic Energy Agency (IAEA), 438 nuclear reactors are operating in the world today.

Radiological Terrorism

This type of terrorism involves detonating a conventional explosive device containing radioactive isotopes with the aim of subsequently dispersing them over a significant area. This category also includes attack scenarios in which radioactive substances are dissolved in water sources. This category of radiation terrorism scenarios is not as powerful in impact as the first category, but is much more likely to be used by terrorists. So-called radiological dispersal devices could be manufactured by packing radioactive materials together with chemical explosives and then detonating the device.

Scenarios for Terrorist Attacks Using Explosives

Because the goal of any terrorist act is to create maximum resonance in society with minimal costs and minimal risk, the use of explosives for terrorist purposes has become widespread. Potential targets of terrorist attacks could include critically important facilities of undoubted interest from the standpoint of inflicting damage and creating significant societal impact.[9]

From the standpoint of the likelihood of technological terrorist attacks, such acts at enterprises using large volumes of flammable substances in their technological processes (gas stations, compressed gas facilities, oil refineries, chemical plants, and so forth) represent a serious potential danger. If explosives are detonated at enterprises using explosive or flammable substances, the following attack scenario is possible: (a) release and dispersal of large volumes of flammable substances, (b) their mixture with air in the necessary proportions and formation of an explosive cloud, and (c) its subsequent explosion. The detonation of explosive clouds over a city could lead to significant destruction and fatalities. Facilities using poisonous substances must be considered as a separate category. Significant destruction at such sites is capable of releasing into the atmosphere a large volume of poisonous substances circulating in the facility's systems, which could contaminate large areas of the city. A separate target for potential technological attacks by terrorists could be a city's natural gas system (gas distribution points, stations, pipelines, underground facilities, and even individual apartments with gas appliances).

Explosive transformation is generally classified in one of two categories deflagration and detonation which are differentiated by the dynamics of the explosive load. The main impact factors from a detonation explosion are an atmospheric blast wave characterized by excess pressure and the force of the compression wave and a fireball created by extremely hot combustion products. The main impact factors from a deflagration explosion are (1) a compression wave characterized by maximal excess pressure, (2) dynamic pressure, (3) wind effects that can substantially exceed centrifugal load, (4) and a fireball of extremely hot combustion products.

MODELING TERRORIST THREATS AND TERRORIST ATTACK SCENARIOS

As noted previously, the distinguishing characteristics of technological terrorism scenarios and impact factors are shaped by the capacity of terrorists for deliberately choosing the means, place, and time for the attack. This choice is based on a rational assessment of (1) the vulnerability of the given target to various attack scenarios and (2) the magnitude of the damages expected if the various attack scenarios are carried out. The decisions made by the terrorists are based on the minimax principle, which consists of a striving to inflict maximum damage on society while expending the minimum resources and with minimal risk that the organization will be detected and eliminated (that is, a striving to ensure maximum effectiveness for the attack). Here, terrorists are capable of reacting to the actions of antiterrorist forces, drawing lessons from the experience of previous attacks, and using them to correct their actions. Additional difficulties that must be faced in evaluating the likelihood that various terrorist attack scenarios will be carried out are associated with the value system of terrorists (that is, their usefulness function) differing notably from the traditional value system. Their system of motivating principles often is not fully comprehensible even to specialists.

Furthermore, the following characteristics are typical of the issues faced in evaluating terrorist attack threats (impact factors) and scenarios:

- High level of uncertainty due to lack of knowledge of terrorists' intentions, intellectual potential, and organizational-technical resources, the goals they are pursuing, and the value system by which they are guided
- Fragmentary and (often) secret nature of data of various types obtained from various sources, such as statistical information, expert assessments, and operational information obtained from intelligence services
- Dynamic nature of terrorist risks

The mathematical model being developed for evaluating various terrorist attack scenarios for a given target must meet the following requirements:

- The model must facilitate assessments and decision making for situations involving a very high level of uncertainty.
- The model must be multidimensional; that is, it must consider a situation from the standpoint of both terrorists and antiterrorist forces. It must provide for a description of the dynamic interaction of these two sides, each of which is guided by its own strategy and is capable of reacting to its opponent's actions. Furthermore, the model must make it possible to take into account terrorists' capacity for selecting the attack scenario that ensures maximum attack effectiveness. That is, it must include the two-way linkages between the vulnerability of the system to

the given attack scenario (and the expected damage) and the likelihood that this attack scenario will be selected by terrorists.[10]

- The part of the model that characterizes terrorists' situational analysis and decision making (part 1 of the model) must assess terrorists' goals, value system, resources, and intellectual and organizational-technical potential; identify basic scenarios for terrorist attacks against a given target; and must assess the probability that various terrorist attack scenarios will be carried out based on their usefulness function, by which (in the opinion of antiterrorist analysts) terrorists must be guided.
- In addition to providing an assessment of vulnerability of the given target and the effectiveness of its protection systems, the blocks of the model that describe the situation from the antiterrorist standpoint must also use results obtained on the basis of analysis of the terrorist part of the model (particularly the likelihood of various attack scenarios being realized from the viewpoint of the terrorists) to determine the most effective measures for countering the terrorist threat. In this regard, the possibility of interaction among the various forces countering the terrorist threat and of exchanges of information among them must be taken into account.
- The model must be dynamic; that is, it must make it possible to describe the change of parameters of the system (target), the external environment, and the spectrum and intensity of terrorist threats.

Given these requirements, it makes sense to bring to bear the principles of game theory[11] and Bayesian networks,[12] which make it possible to (1) take into account the independent actions and rational behavioral strategies of terrorist and antiterrorist forces; (2) assess situations characterized by high levels of uncertainty; and (3) account for information obtained from various sources (including information received periodically on the status of particular variable models), thus making it possible to obtain detailed inductive assessments of the likely accuracy of the predictions of other variable models.

Scientific methodological aspects and applied developments have become the focus of joint analysis within the framework of a program for countering technological terrorism being carried out jointly by the Russian Academy of Sciences and the U.S. National Academy of Sciences[13] and under the Science for Peace Program of the North Atlantic Treaty Organization.[14]

Figure 7-2 presents a three-sided model that facilitates assessment of terrorist attack scenarios and counteractions by antiterrorist forces. The model consists of three graphs. Graph 1 is a diagram of influence describing the situation involved in making decisions to select an attack scenario from the standpoint of a terrorist organization. This diagram is compiled by analysts at the security service of a target facility, who, in their attempt to consider things from the terrorists' position (playing the role of the enemy), strive to assign values for expected usefulness for the terrorists if various attack scenarios were to be carried out. The values arrived

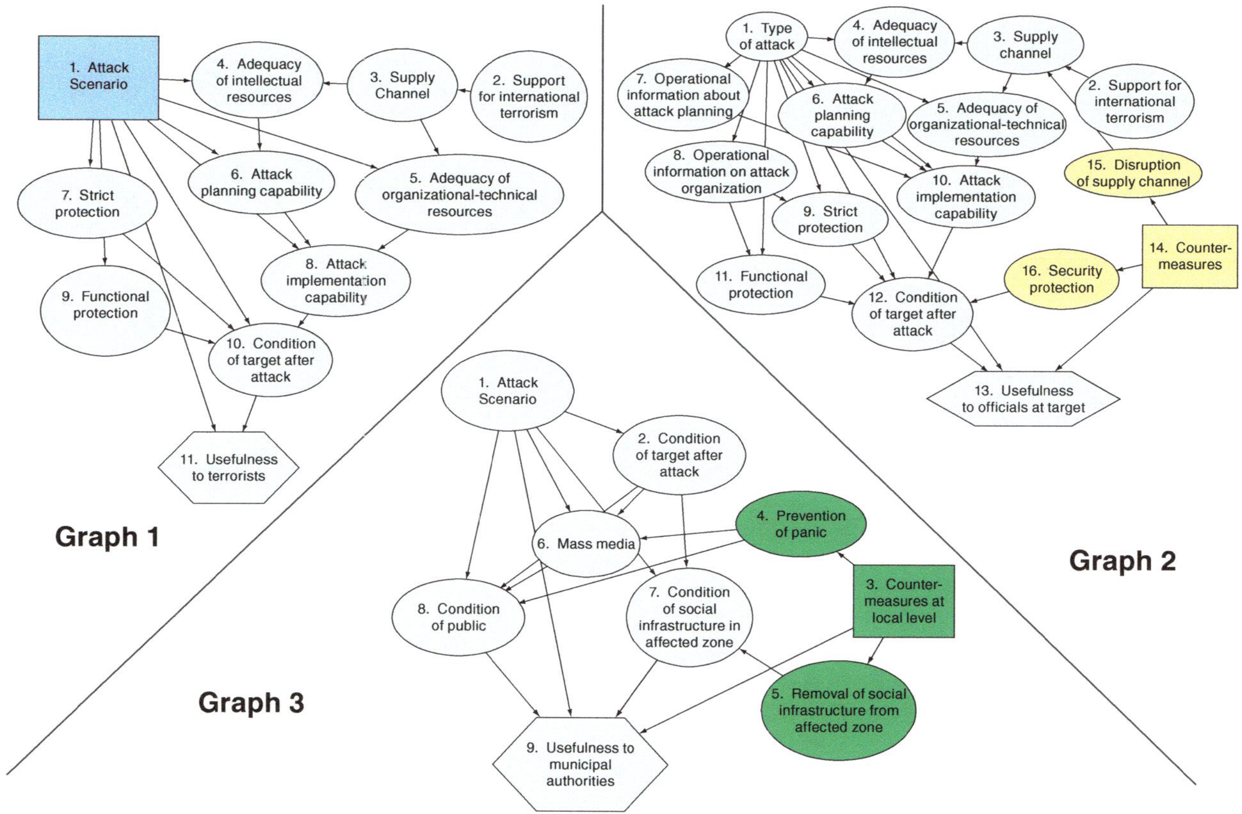

FIGURE 7-2 Three-sided terrorism risk assessment model.

at then make it possible to assess the probability of the various attack scenarios being realized. These probabilities are used in constructing graphs 2 and 3, which characterize the corresponding process of making decisions to select measures to counter the terrorist threat at the level of the security service of the given facility (Graph 2) and at the level of the municipal authorities in the area where the facility is located (Graph 3). It should be kept in mind that facility officials and the municipal authorities can exchange information and coordinate their efforts; that is, they are allies in the game.

The similarities and differences between the graphs show that they describe the same question but from different positions. For instance, the differences between the graphs reflect the varying level of uncertainty regarding the condition of specific elements (the condition of the same element—for example, the resources of the terrorist organization could be known for certain by the terrorists but viewed by antiterrorist forces as a random value). Assessments of the probability links between task variables (that is, the tables of conditional probabilities for the three graphs) also differ accordingly. In addition, some task parameters may not be considered at all by one side but, at the same time, could be very important to the other. Fundamental differences are also noted in the usefulness elements of each graph, inasmuch as the usefulness functions for terrorists, facility officials, and the municipal authorities may take completely different factors into account. Terrorists, for example, may be oriented primarily toward infliction of the initial blow and on the expenditures necessary for carrying out the attack, while for facility officials the usefulness function must also include secondary damage and the cost of implementing various protective measures. The usefulness function for the municipal authorities must first take into account the damage inflicted on the public and the local infrastructure in areas near the target facility.

NOTES

1. Frolov, K., and G. Baecher. 2006. Protection of the Civilian Infrastructure from Acts of Terrorism. Dordrecht, The Netherlands: Springer, 252 pp.

Pate-Cornell, E. 2002. Probabilistic modeling of terrorist threats: A systems analysis approach to setting priorities among countermeasures. Military Operations Research 7(4):5-20.

Woo, G. 2004. Quantitative terrorism risk assessment. Available online at *www.rms.com/NewsPress/Quantitative_Terrorism_Risk_Assessment.pdf*. Accessed April 11, 2008.

2. *Technogenic* is used to refer to phenomena arising as a result of the development or deployment of technology.

3. Makhutov, N. A., and D. O. Reznikov. 2007. Use of Bayesian networks to assess terrorist risks and select an optimal strategy for countering the terrorist threat. Problems of Security and Extreme Situations 5:43-63.

Pate-Cornell. Probabilistic modeling of terrorist threats.

4. Fortov, V. E. 2004. Study of electromagnetic impacts in terrorist and antiterrorist actions. Pp. 228-238 in Proceedings of a Scientific-Practical Conference. Moscow: Kombitell.

5. Barsukov, V. 2000. Protecting computer systems from powerful destructive effects. Jet Info Information Bulletin 2(81):8-17. Available online at *www.jetinfo.ru/2000* (in Russian).

Vasenin, V. A., and A. V. Galatenko. 2002. Computer terrorism and Internet security problems. Pp. 211-225 in High-Impact Terrorism: Proceedings of a Russian-American Workshop. Kirov, Russia: Vyatka. [Pp. 183-197 in the original English version by the same title, published in 2002, Washington, D.C.: The National Academies Press.].

Branscomb, L. 2003. Cyberattacks as an amplifier in terrorist strategy. Pp. 93-96 in Terrorism: Reducing Vulnerabilities and Improving Responses: U.S.-Russian Workshop Proceedings. Washington, D.C.: The National Academies Press.

6. Morenkov, O. S. 2002. Bioterrorism: A view from the side. Pp. 131-141 in High-Impact Terrorism: Proceedings of a Russian-American Workshop. Kirov, Russia: Vyatka. [Pp. 106-113 in the original English version by the same title, published in 2002, Washington, D.C.: The National Academies Press.].

McGeorge, J. 2001. An analysis of 404 nonmilitary incidents involving either chemical or biological agents. P. 53 in Abstract Book of the World Congress on Chemical and Biological Terrorism, Dubrovnik, Croatia, April 22-27, 2001.

7. Ibid.

8. Aratyunyan, R. V., V. Belikov, et al. 1999. Models for the spread of radioactive contamination in the environment. RAS Power Engineering News 1:61-96.

Hecker, S. 2002. Nuclear terrorism. Pp. 176-184 in High-Impact Terrorism: Proceedings of a Russian-American Workshop. Kirov, Russia: Vyatka. [Pp. 149-155 in the original English version by the same title, published in 2002, Washington, D.C.: The National Academies Press.]

9. Komarov, A. A. 2004. Questions of protecting the urban infrastructure and the public from explosive technological terrorism and catastrophic explosions. Pp. 79-89 in Proceedings of a Scientific-Practical Conference. Moscow: Kombitell.

Simmons, R. 2002. Terrorism: Explosives threat. Pp. 199-211 in High-Impact Terrorism: Proceedings of a Russian-American Workshop. Kirov, Russia: Vyatka. [Pp. 171-179 in the original English version by the same title, published in 2002, Washington, D.C.: The National Academies Press.]

10. The use of two-sided models describing the terrorist and antiterrorist sides of a conflict is described in detail in Pate-Cornell, Probabilistic modeling of terrorist threats.

11. Hausken, K. 2002. Probabilistic risk analysis and game theory. Risk Analysis 22(1):17-27. McCain, R. Game theory: An introductory sketch. Available online at *william-king.www.drexel.edu/top/eco/game/nash.html.*

Sandler, T., and D. Arce. 2003. Terrorism and game theory. Simulation and Gaming 34(3).

12. Terekhov, S. A. 2003. Introduction to Bayesian networks. Scientific session of the Moscow Engineering-Physics Institute, Fifth All-Russian Scientific Practical Conference, Moscow; Jensen, F. V. An Introduction to Bayesian Networks. 1996. New York: Springer-Verlag.

13. National Research Council Committee on Counterterrorism Challenges for Russia and the United States. 2004. Terrorism: Reducing Vulnerabilities and Improving Responses: U.S.-Russian Workshop Proceedings. Washington, D.C.: The National Academies Press.

14. Frolov and Baecher. Protection of the Civilian Infrastructure.

Additional background materials not specifically cited:

Petrov, V. P., D. O. Reznikov, V. I. Kuksova, and Ye. F. Dubinin. 2007. Terrorist risk assessment and decision-making on the expediency of building a protection system against terrorist actions. Pp. 89-105 in Problems of Security and Emergency Situations, vol. 1.

Tucker, J. 2002. Chemical terrorism: Assessing threats and responses. Pp. 141-165 in High-Impact Terrorism: Proceedings of a Russian-American Workshop. Kirov, Russia: Vyatka. [Pp. 117-133 in the original English version by the same title, published in 2002, Washington, D.C.: National Academies Press.]

Frolov, K. V., and N. A. Makhutov. 2004. Technological terrorism and methods of countering terrorist threats. Pp. 228-238 in Proceedings of a Scientific-Practical Conference. Moscow: Kombitell.

8

Activities of the Russian Federal Medical-Biological Agency Related to Radiation, Chemical, and Biological Security*

Vladimir V. Romanov, Deputy Head of the Russian Federal Medical-Biological Agency (FMBA) and Chief State Sanitary Physician for Organizations and Territories Served by FMBA

Organizationally, the Russian Federal Medical-Biological Agency (FMBA) is a unified complex of clinical, prophylactic, sanitary, antiepidemic, and research organizations whose activities are aimed at improving working conditions for personnel in especially hazardous industries and detecting and eliminating the effects of harmful physical, chemical, and biological factors on the health of workers and the public living near dangerous facilities. The agency includes 92 clinical-prophylactic facilities (central medical-sanitary units, medical-sanitary units, and clinical hospitals), 19 scientific research institutes, 42 regional (inter-regional) offices, and 63 hygiene and epidemiology centers.

Celebrating its 60th anniversary in 2007, FMBA is the successor of the Third Main Administration of the USSR Ministry of Health, which was organized in 1947 to provide medical and sanitary-hygiene support for efforts to create nuclear weapons. The administration was later assigned tasks related to monitoring working conditions for chemical weapons industry workers and for handling disease-prevention measures both for manned space flights and for organizations working with pathogenic microorganisms in hazard classes 1-4.

---

*Translated from the Russian by Kelly Robbins.

In accordance with existing Russian Federation legislation, FMBA is responsible for medical-sanitary support functions and state sanitary-epidemiological monitoring for organizations in certain industries in which working conditions are particularly hazardous and for the population in certain areas (Decree of the Russian Federation President No. 1304, On the Federal Medical-Biological Agency, dated October 11, 2004; Russian Federation Government Resolution No. 789, Issues Regarding the Federal Medical-Biological Agency, dated December 15, 2004; and Russian Federation Government Resolution No. 206, On the Federation Medical-Biological Agency, dated April 11, 2005). It also performs state regulatory functions related to the use of nuclear power (Russian Federation Government Resolution No. 412, On Federal Executive Branch Agencies Involved in State Management of the Use of Nuclear Power and State Regulation of Safety in the Use of Nuclear Power, dated July 3, 2006). There is no comparable organization in the United States that focuses on very hazardous environments at nuclear, chemical, and biological facilities.

According to Russian Federation government directives (No. 1156-r of August 21, 2006, and No. 1745-r of December 16, 2006), the list of entities served by FMBA includes all of the main radiation-, chemical-, and biological-hazard organizations operating under the auspices of the Federal Atomic Energy Agency, the Federal Industrial Agency, the Federal Oversight Service for the Protection of Consumer Rights and Human Welfare, and other executive branch agencies of the federal government and the various jurisdictions in which those organizations are located.

FMBA carries out its activities both directly and through its subsidiary local monitoring offices and organizations. The local offices and FMBA hygiene and epidemiology centers are part of the unified system of agencies and institutions responsible for state sanitary-epidemiological oversight in the Russian Federation. Policies and procedures governing their activities are set forth in Russian Federation Government Resolution No. 569, Statute on the Provision of State Sanitary Epidemiological Oversight in the Russian Federation, dated September 15, 2005.

The scientific research institutes under FMBA's auspices provide scientific support for the activities of the agency's practical health care institutions, local offices, and hygiene and epidemiology centers. They study the health status of assigned populations and provide state sanitary-epidemiological oversight in the development of regulatory-legal acts on monitoring of organizations presenting radiation, chemical, and other hazards.

Furthermore, FMBA's scientific research institutes have produced fundamental results in studying the effects of ionizing radiation on the human body, radiobiology, and radiation medicine and hygiene and in developing medical preparations that protect against radiation and chemical impacts and individual gear and devices that protect the respiratory systems and skin of workers at radiation-hazard facilities. The institutes have also made progress in the area of biological instrument manufacturing, new-generation vaccine development, and research on the immune status of workers at hazardous facilities, among other developments.

The State Science Center—Institute of Biophysics includes Russia's only clinical department specializing in the treatment of radiation-related conditions (the Occupational Pathology Department) and also features an emergency medical dosimetry center. The center was created as an emergency response unit. It is functionally included in the Federal Atomic Energy Agency's Crisis Center and is responsible for providing support for the activities of FMBA local offices and institutions in the assessment of the radiation situation in areas affected by radiation accidents and in management decision making on emergency response measures by FMBA subunits.

FMBA's accumulated expertise and the many research developments it has made in the areas of radiation, chemical, and biological safety must undoubtedly be used to protect the population of the Russian Federation from the current level of terrorist threats. FMBA is open to cooperation and is prepared to work within the framework of joint U.S.-Russian research projects to prevent threats of high-technology terrorism.

9

Disease Surveillance and International Biosecurity*

David R. Franz, Midwest Research Institute

Biological Security and Human Security: Biological security is fundamental to human security; human security is fundamental to stability in this ever smaller and more connected world. The perceived threats to our biological security today are described by a broad spectrum of risks worldwide. Where each of us finds ourselves on that spectrum depends to a great extent in which part of the world we reside. The enormous impact of chronic disease (cancer, heart disease, diabetes, and so forth) and communicable and contagious disease (HIV/AIDS, malaria, tuberculosis, and hepatitis), the potentially very significant impact of emerging diseases such as Severe Acute Respiratory Syndrome (SARS) and highly pathogenic influenza, the potentially large impact but low likelihood of bioterrorism, and the emerging concern for the exploitation of dual-use biotechnologies to cause harm all receive different attention and concern regionally across the globe. Therefore, biological risk perception is related to technological advancement, the state of public health and political factors within a given region

*This report expands on and updates a paper by Franz, David R. 2007. Species-neutral disease surveillance: A foundation of risk assessment and communication. Pp. 93-99 in Risk Assessment and Risk Communication Strategies in Bioterrorism Preparedness, M. Green, J. Zenilman, D. Cohen, I. Wiser, and R. Balicer, eds. Dordrecht: Springer.

or country. Where a country or region finds itself on the spectrum will change over time with these factors as well. However, because of rapid changes in transportation and movement of humans and animals, we are clearly all impacted by any major event or outbreak anywhere in the world.

Global Emerging Infectious Disease Risk: In December 1979 the World Health Organization declared smallpox eradicated from the globe. After this enormous victory over disease, accomplished by a serious joint effort between Russian, U.S., and collaborating public health leaders from around the globe, the infectious disease community felt good about its ability to control biological risks. Some even speculated that we would now move on from smallpox to polio, malaria, and other important scourges, until we made the world free from infectious disease. It was not to be. In 1992 the Institute of Medicine of the U.S. National Academies published a report authored by Professors Lederberg, Shope, and Oaks entitled *Emerging Infections: Microbial Threats to Health in the United States*. The report, while making several important recommendations for the future, stated, "Disease-causing microbes have threatened human health for centuries. The Institute of Medicine's Committee on Emerging Microbial Threats to Health believes that this threat will continue and may even intensify in coming years."[1] Those words proved to be prophetic.

An Important Lesson in the United States: In late June 1999 an unusual number of dead birds were reported in the borough of Queens, New York City. Six to 8 weeks later, an unusual number of human cases of encephalitis were noted in hospitals in the area. The human disease was soon diagnosed as St. Louis encephalitis, a mosquito-borne viral encephalitis, the causative agent of which does **not** kill birds. Approximately 2 weeks after the first human cases, the "St. Louis" outbreak was announced and mosquito control was begun. Then, 2 or 3 weeks later, animal disease data and human disease data were integrated, and the true causative agent, West Nile virus, was implicated in both the bird and the human deaths. We will never know if, or exactly how many, lives and dollars might have been saved by knowing 6 weeks earlier that a new, deadly zoonotic arbovirus had been introduced to North America, but experts agree days—and sometimes even hours—make a real difference when dealing with infectious outbreaks. Had we been thinking in terms of disease—wherever it occurs—rather than just human disease, we might have done better. The origin of the virus in North American birds or mosquitoes is unknown, but there is little doubt that it came from outside the borders of the United States.

Discovering Disease Early: Whether a disease is introduced naturally, accidentally, or intentionally, one of the most important factors is discovering it as early as possible and understanding its spread through the population. Because many diseases of concern to humans are first seen in animals—West Nile encephalitis, SARS, monkey pox, and H5N1 influenza are recent examples—it is critical that we seek to discover disease in the species of origin. Finding evidence of a zoonotic disease first in animals will very likely allow preventive or prophylactic actions to be taken to protect the human population. The concept of species-neutral disease surveillance acknowledges that we must look for "disease" wherever it is found, not "human disease" and "animal disease." Finally, we live in a much smaller world than we did just a few decades ago. Transportation and travel are such that an outbreak in one part of the world today can impact humans or animals on the other side of the globe tomorrow. Therefore, there is great benefit in discovering an outbreak (1) as early as possible, (2) in the host species of origin, and (3) in the region of origin.

Traditionally, disease surveillance in most countries has involved a Ministry of Health network that monitors human disease and a Ministry of Agriculture system that monitors disease in animals. Even in the twenty-first century—in technologically advanced democracies—it is not uncommon for these two activities to go on in parallel, sometimes discovering the same outbreak in their own species of responsibility, without effective communication between them. During the U.S. introductions of West Nile virus (1999) and monkey pox (2003), for example, communication between the animal health and human health professionals was less than adequate. Likewise, nations have been reticent to tell other nations that they have discovered a disease outbreak on their soil, fearing negative travel, trade, and economic consequences. The same attitudes and practices have been the norm in many nations and regions throughout the era of modern public health. Knowing that approximately 75 percent of emerging infectious diseases and many of those agents traditionally selected for use as weapons are zoonotic, it only makes sense that we must integrate our disease surveillance efforts. The situational awareness that an integrated disease surveillance program provides must be a key component of our preparation and response if we are to be prepared to minimize loss of life and economic impact when disease outbreaks occur.

One Health: Supporting species-neutral disease surveillance, there has been a recent revival of the concept of "One Health," encouraging increased communication between human, animal, and plant "health-care providers."[2] In the summer of 2007 the American Medical Association and the American Veterinary Medical Association formally adopted resolutions to work together to enhance "collaboration between human and veterinary medical professions in medical education, clinical care, public health, and biomedical research."[3] This acknowledgment and initiative by these two prominent professional organizations should provide long-needed impetus to strengthen public and animal health in the United States. International collaboration on One Health would further improve the disease surveillance and health situation worldwide while enhancing understanding and facilitating communication and progress in the life sciences.

International Compact for Infectious Disease: Dr. Harvey Rubin of the University of Pennsylvania in the United States, working with international colleagues, has proposed an International Compact for Infectious Disease to deal with pandemic, epidemic, and endemic infectious diseases that threaten personal, national, and international security. The compact would include four pillars, the first of which is focused on disease surveillance:

1. Establishment, maintenance, and monitoring of international standards for surveillance and reporting of infectious diseases using advanced information technology to ensure timeliness, interoperability, and security
2. Establishment, maintenance, and monitoring of international standards for best laboratory practices
3. Expansion of capabilities for the production of vaccines and therapeutics expressly for emerging and reemerging infections
4. Establishment, maintenance, and monitoring of a network of international research centers for microbial threats

The idea of the proposed compact[4] is currently being presented to medical, scientific, and public health organizations and audiences worldwide in an effort to refine the concept and seek support for its eventual adoption. This approach and the interest it has generated is further evidence that the need for global responses to global infectious disease threats is gaining visibility and support worldwide.

Technical Tools for Early Discovery of an Outbreak: Both the apparently increasing frequency with which we have faced newly emerging disease in recent years and the intentional anthrax attacks experienced in the United States in 2001 have motivated some nations to take disease surveillance more seriously.

Data Mining: One form of surveillance, data mining, attempts to cast a very wide net to discover the human response to a disease outbreak: ambulance calls, over-the-counter medications purchased, emergency room visits, and disease-related information sought. Others, such as ProMed,[5] allow users around the globe to input disease-related notices. Yet another model, ARGUS,[6] employs a team of humans with skill in more than 30 languages to actively monitor Web-based news reports looking for early reporting of unusual disease or outbreaks worldwide. While these passive and active programs, exploiting the power of the Web, might highlight trends and provide enough information to help us connect the dots, the signal-to-noise problem with some of the systems has limited their utility. Efforts to develop and implement automated syndromic surveillance systems continue as we seek more efficient ways to provide situational awareness regarding new outbreaks and disease prevalence.[7]

Clinician-Driven Syndromic Surveillance: Amazingly, we believe that the index case of inhalational anthrax was discovered by an astute clinician following the mailing of *B. anthracis*-laced letters in the fall of 2001. There are several clinician-driven surveillance systems undergoing testing today.[8] The greatest challenges of implementing an effective, clinician-driven system are probably (1) difficulty in down-selecting to implement just the right system, (2) failure to have widespread connectivity throughout the various venues in which clinicians see patients, (3) the time it takes from the clinicians' busy schedule to input required data, and (4) few clinician-driven disease surveillance systems being tested today integrate human and animal disease data. There are efforts under way within the Department of Homeland Security[9] to deal with the first problem, by developing a higher-level integrating information system that can take inputs from numerous disparate collection systems. As with data mining, there are trade-offs in implementing clinician-based surveillance systems. While we will certainly discover evidence of disease wherever systems are in place, our discovery will come when there is already disease in the population. On the other hand, it may be easier to justify long-term maintenance of clinician-driven surveillance because these systems will be valuable across the entire biothreat spectrum. As preclinical diagnostics improve and find a place in the clinic, or even the home or workplace, we may be able to move ever closer to the index case and even the index exposure, the elusive goal of environmental monitoring.

Environmental Monitoring has been adapted for early warning of a biological terrorism event in the United States and other countries. The advantage of environmental sensors is that, if located at the right place, programmed to sample ambient air at the right time, and designed to identify the agent being used, they could warn us of disease-causing organisms in the air we breathe even before our citizens become ill. These systems could be placed in the top 40-50 population-dense centers, operating 24 hours a day all year long, for high tens of millions of dollars or low hundreds of millions. As currently configured, systems of such sensors will likely not warn us of a naturally emerging disease outbreak or even aerosolized novel pathogens or different strains. Both the concept of dispersed environmental sensors and passive data mining may be hard to justify for the many years that the low-likelihood threat of bioterrorism may exist.

The Tools Are Getting Better: Although there have been various efforts to improve disease surveillance in the United States, one that stands out by combining true clinician-based syndromic surveillance across species is the Syndromic Reporting Information System (SYRIS).[10] This system allows simple and direct Web-based input by physicians or veterinarians, central monitoring and collation by public health authorities, and—very importantly—rapid feedback to the clinician regarding regional disease reporting. The human portion is based on six syndromes: (1) influenza-like illness, (2) acute hepatitis, (3) acute respiratory distress syndrome, (4) fever with skin rash, (5) fever with central nervous system findings, and (6) fever with severe diarrhea. The nine veterinary syndromes are: (1) lymphadenopathy with fever, (2) severe diarrhea, (3) off feed/wasting, (4) neurological/lameness, (5) vesicular lesions, (6) pneumonitis, (7) downer animal, (8) drooling/slobbering, and (9) dead. Within 30 seconds after entering as little as a syndrome and a zip code (mail code), the clinician receives a map showing the reported syndrome and any related information from other clinicians, human and veterinary, in the area or region. This system facilitates an awareness that is unprecedented in both time and species covered. SYRIS is being used in part of the state of Texas and being tested in several locations within the United States.

Although available anywhere the Web can be accessed, SYRIS is not currently being implemented outside the United States. There are, of course, many new technical developments that improve our ability to identify and characterize disease-causing agents, but we still lack a universal nationwide system for early identification of disease in the population and a codified response protocol.

We have the tools. Do we have the will? Exactly how to accomplish the important task of very early awareness and response to naturally emerging and intentional disease is not yet clear, but its importance is beyond question. We must watch "that spot where the animals, humans, and bugs collide." We now have the technical tools and know-how to implement the necessary public health infrastructure for surveillance and response nearly anywhere around the globe. The world has become too small and the potential for harm too great to stand idle. Technology allows implementation today; we must not let politics or borders stand in the way. Working together, across national boundaries on one of the most challenging and important human security issues of our time, will not only protect our citizens from natural disease but also contribute to building understanding and even trust that will reduce the likelihood that intentional outbreaks will negatively impact any of our populations.

NOTES

1. Lederberg, J., R. E. Shope, and S. C. Oaks, Jr., eds. 1992. Emerging Infections: Microbial Threats to Health in the United States. Washington, D.C.: National Academy Press, p. 1. Available online at *www.nap.edu/catalog/2008.html#toc*. Accessed December 3, 2007.
2. Zinsstag, J., et al. 2005. Potential of cooperation between human and animal health to strengthen health systems. Lancet 366: 2142-2145; Gibbs, E. P. J. 2005. Emerging zoonotic epidemics in the interconnected global community. Veterinary Record 157: 673-679.
3. American Medical Association House of Delegates, Resolution: 530 (A-07). Available online at *www.ama-assn.org/ama1/pub/upload/mm/467/530.doc*. Accessed December 3, 2007.
4. Rubin, H., et al. 2004. The New Arms Race: Making a Case for an International Compact for Infectious Disease. Philadelphia: ISTAR and University of Pennsylvania. Available online at *www.istar.upenn.edu/compact/downloads.html*. Accessed December 6, 2007.
5. See *www.promedmail.org*.
6. See *biodefense.georgetown.edu/projects/argus.aspx*.
7. See *www.syndromic.org*.
8. See *www.zelicoff.com/SMLR/SyndromicSurveillancePage/defaultSyndrome.htm*.
9. See *www.dhs.gov/xlibrary/assets/mgmt/e300-prep-nbis2008.pdf*.
10. Available online at *www.zelicoff.com/SMLR/SyndromicSurveillancePage/defaultSyndrome.htm*.

10

Emerging Viral Infections in the Asian Part of Russia

Sergei V. Netesov, Federal State Research Institution—State Research Center of Virology and Biotechnology Vector and Novosibirsk State University, and Natalya A. Markovich, Federal State Research Institution—State Research Center of Virology and Biotechnology Vector

The so-called emerging infections are primarily the result of increased human activities such as international and domestic trade, tourism, industrialization and its consequences, and, to a lesser extent, climate change, which are detailed in this paper. The stages of emergence and spread of highly pathogenic subtype H5 avian influenza virus over the territory of Russia in 2005-2007 are considered, as well as the corresponding measures of its control. It is now well known that the mortality rate resulting from emerging infections is considerably higher than that caused by bioterrorism. At the same time, bioterrorists may use emerging infectious agents for bioterrorist acts. Therefore, emerging and reemerging infectious diseases are of substantial interest in the biosecurity community because some of these agents can be used for intentional attacks. In addition, natural outbreaks can help highlight vulnerabilities and gaps in public health or agricultural response capabilities. Therefore, it makes sense to intensify development of measures to control emerging infections, which will also enhance the struggle against bioterrorism.

Since the 1980s, physicians and specialists have encountered the emergence of new infectious diseases—emerging infections—with ever-increasing frequency. The new emerging infections appear once every 2 or 3 years. Each infection has its own specific features; therefore, each requires special attention from scientists and public health care practitioners.

Seven main reasons underlie the emergence of new infections:

1. Transfer of infections from one region of the world to another by migratory birds
2. Human colonization of new territories inhabited by previously unknown animals or insects
3. Industrial breeding of animals, particularly new animal species, or introduction of new species of pets
4. Introduction of animals to new territories where they have not previously lived
5. Global warming and the subsequent invasion of new animal and insect species
6. Creation of new conditions for reproduction of animals and insects as a consequence of human activities
7. Adoption of new technologies, which not only improves human life but also creates new conditions for the reproduction of pathogenic microorganisms

Let us consider each of these issues individually.

TRANSFER OF INFECTIONS BY MIGRATORY BIRDS (EXAMPLE: H5N1 INFLUENZA VIRUS)

The map in Figure 10-1 illustrates the epizootic caused by avian influenza virus in western Siberia in the summer of 2005. The first mass mortality event, initially affecting wild birds and subsequently domesticated species, was recorded by the Federal Agency for Veterinary and Phytosanitary Surveillance (Rosselkhoznadzor) in the village of Suzdalka, Dovolnoye Region, Novosibirsk Oblast (Shestopalov et al., 2006; Evseenko et al., 2006; and Lipatov et al., 2007). After the regional Rosselkhoznadzor office received the report from this village about the deaths of wild birds, it notified other regions of Novosibirsk Oblast and neighboring jurisdictions. Veterinarians began detecting disease in wild birds and later in domesticated birds at many sites in western Siberia and reported their findings to the local Rosselkhoznadzor offices. In particular, analogous epizootics were recorded in July-August 2005 in wild birds and subsequently in domesticated species in Altai Krai and Novosibirsk, Tomsk, Kemerovo, Omsk, and Kurgan oblasts, as well as in Pavlodar Oblast in Kazakhstan (Lipatov et al., 2007).

Note that the threat to domesticated birds in individual and commercial farms was very serious, as these regions are known for mass poultry breeding on both private and industrial scales. Physical security is practically absent on individual farms but is sufficient on most commercial farms, which are equipped with ventilation tubes and doors as well as mesh-covered windows to prevent wild and domestic birds from mixing. In addition, at commercial farms grain is heat treated before feeding, and personnel and their clothing are disinfected at build-

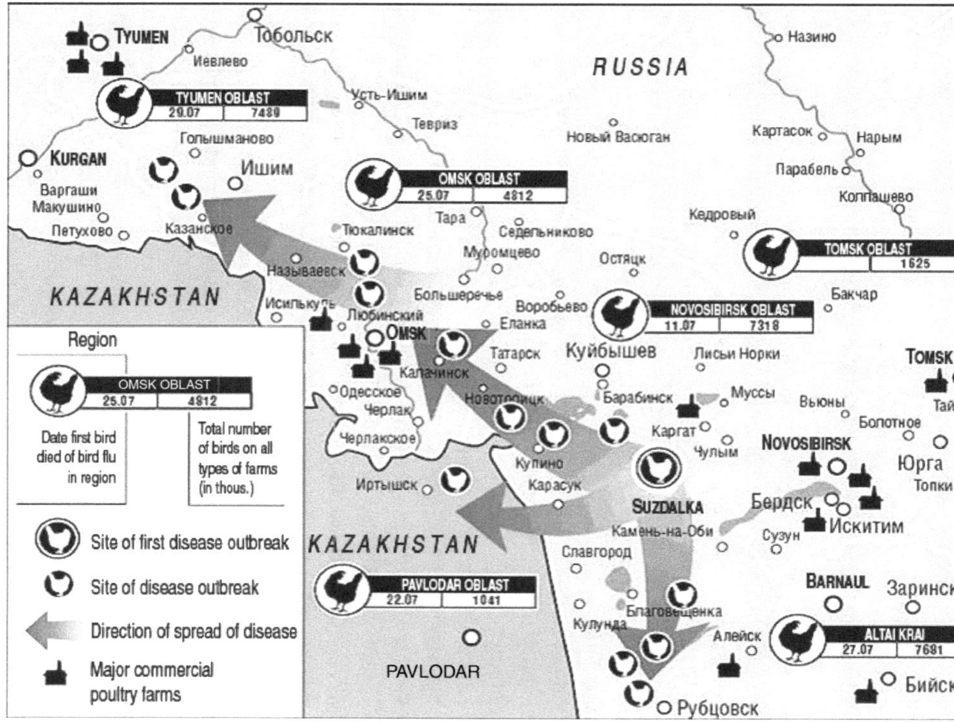

FIGURE 10-1 Map of epizootics caused by H5N1-subtype avian influenza virus in Russia in summer 2005.
NOTE: In the black boxes, the names of regions, amount of both individual and industrial domesticated birds, and dates of first notification made by local (county) Rosselkhoznadzor officers are shown.

ing entrances and exits. The poultry stock in individual and commercial farms in Novosibirsk Oblast amounts to approximately 7.3 million; in Omsk Oblast, 4.8 million; in Tyumen Oblast, almost 7.5 million; and in Altai Krai, nearly 7.7 million birds. One of the most likely causes of mortality among domesticated birds was avian influenza virus, which, as is known from the events of 2003-2004 in Southeast Asia, represents a tremendous threat for poultry farming. Thus, it was clear that the most serious measures were required to control this disease, even not taking into account the potential threat of human morbidity and mortality.

The sequence of events for diagnosis of the disease and study of the properties of avian influenza virus strains in Russia in July 2006 was as follows:

- July 11, 2006: A gamekeeper from the village of Suzdalka and a veterinary officer from Dovolnoye Region reported to the regional office of Rosselkhoznadzor about a mass mortality of wild birds on Suzdalka Lake.
- July 15, 2006: The first recording of the mass mortality event among domesticated birds in the village of Suzdalka, Novosibirsk Oblast, was made; the first team from the State Research Center of Virology and Biotechnology Vector was sent to Suzdalka.
- July 17-18, 2006: A sampling of organs and feces from domesticated birds was taken by the Vector team, and the samples were delivered to Vector; assaying of samples began.
- July 20, 2006: The first results of analysis (identification of the pathogen as the H5 subtype of the avian influenza virus) were reported to the governor of Novosibirsk Oblast and the regional and central offices of the Federal Monitoring Service for Consumers' Rights and Welfare (Rospotrebnadzor).
- July 22, 2006: The results of further analysis (genotype H5N1 and high pathogenicity for chicks) were reported to the governor and offices of Rosselkhoznadzor and Rospotrebnadzor.
- July 24, 2006: Complete nucleotide sequences of hemagglutinin (HA) and neuraminidase (NA) genes of the virus were determined and their phylogenetic similarity to influenza virus strains isolated in China in April-May 2005 from birds on Qinghai Lake was demonstrated; data on potential pathogenicity for humans were obtained based on molecular genetic characteristics (Evseenko et al., 2006).
- July 24, 2006: A specialized commission was organized on order of the governor of Novosibirsk Oblast for control of the epizootic (control measures included exterminating infected birds, disseminating information about the epizootic to the public, and preventing the spread of the epizootic).

The five main properties of the Suzdalka influenza virus strain isolated at Vector are listed below (these data were reported on July 24, 2006, to the governor of Novosibirsk Oblast and territorial and central offices of Rospotrebnadzor):

1. According to serological and genetic data, the isolated strain was of the H5N1 subtype.
2. The hemagglutinin cleavage site of this strain contained six positively charged amino acids, thus indicating its potential pathogenicity for humans.
3. Analysis of the nucleotide sequence of the M2 gene of this strain demonstrated amantadine sensitivity.
4. Phylogenetic analysis of the HA and NA gene sequences demonstrated that this strain was most similar to strains isolated in May 2005 from dead wild migratory birds on Qinghai Lake in central China.
5. Analysis of intravenous pathogenicity in chicks demonstrated that this strain displayed the highest pathogenicity index, IVPI = 3, meaning one-half of the infected chicks died during the first 24 hours after infection (Evseenko et al., 2006; L'vov et al., 2006; and Lipatov et al., 2007).

During July 2005, this virus was detected in the majority of oblasts in the southern part of western Siberia and in several oblasts of Kazakhstan, having caused extensive epizootics with mass mortality of wild fowl and domesticated birds in individual open-type farms. Analogous outbreaks were recorded during 2005 in northern Mongolia and northeastern China, that is, in the particular territories crossed by migratory flyways from China to Kazakhstan and Russia. The epizootics in Novosibirsk Oblast and other oblasts of the southern part of western Siberia were virtually stopped by August 2005 by exterminating the sick birds and their avian contacts in the villages and subsequently disinfecting the farmsteads and preventing domesticated birds from coming into contact with wild aquatic birds.

It was clear that this outbreak could be repeated in the fall, as the migratory birds from northern Siberia would cross southern Siberia on their way to wintering sites. Therefore, to prepare an avian influenza forecast for the fall of 2006, the team of experts from Vector, the Institute of Animal Systematics and Ecology, Siberian Branch of the Russian Academy of Sciences, and the regional Rosselkhoznadzor office constructed a map of autumn flyways for the migratory birds of western Siberia (see Figure 10-2). Based on this map, it was assumed that the H5N1-subtype influenza virus could be brought from the northern part of western Siberia back to southern Siberia as well as to China and Mongolia. The virus could also be transferred to Kazakhstan, Uzbekistan, Turkmenistan, and the European part of Russia, and from these regions to western European countries. In fact, it actually reached eastern and some central European countries. In Turkey and Romania it led to considerable decreases in or even bans of poultry exports, which caused substantial financial losses.

This was in fact what happened. Influenza outbreaks in the fall of 2005 were recorded in Tula, Moscow, Chelyabinsk, Tambov, Novosibirsk, Omsk, Kurgan, and Tyumen oblasts and in Altai Krai. In addition, in Kurgan Oblast the virus first appeared in a large commercial poultry farm. Consequently, the entire stock

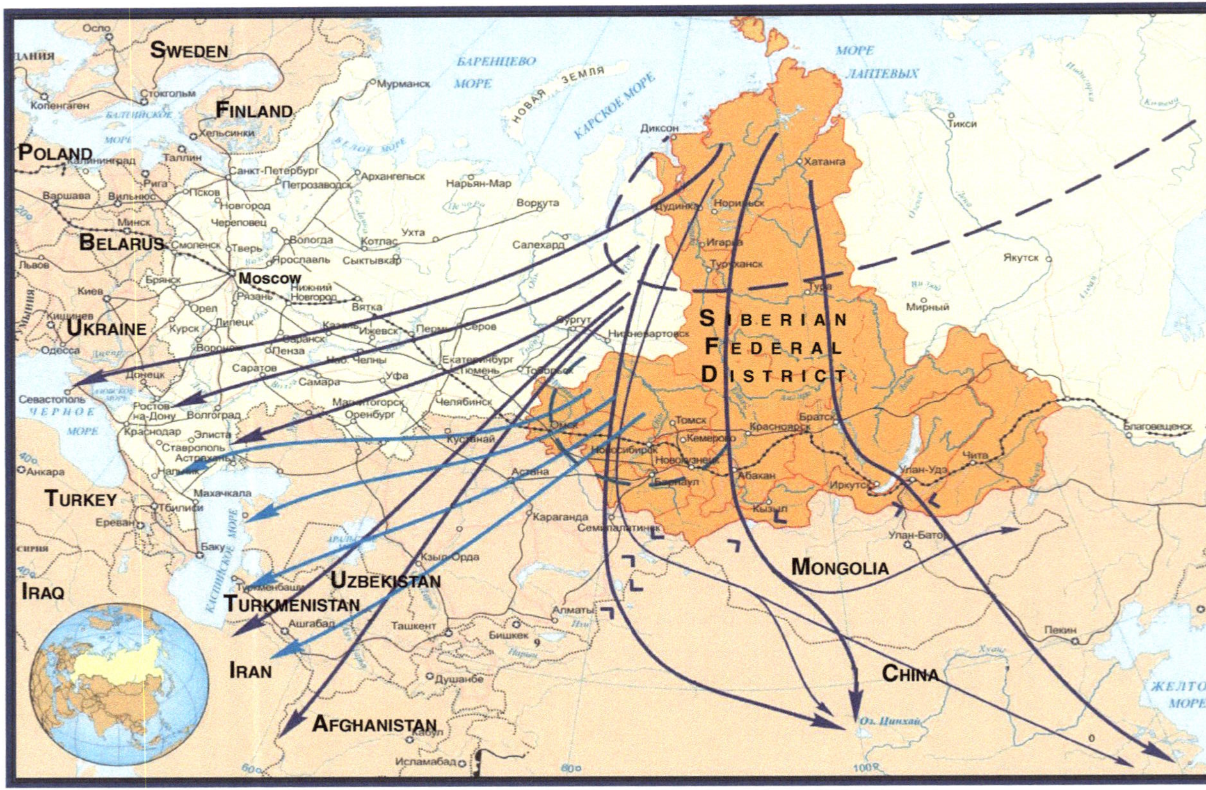

FIGURE 10-2 Flyways of the fall migration of migratory birds crossing the territory of Siberian Federal District.

of 450,000 hens and chicks was exterminated. This virus was also transferred to European countries: Epizootics occurred in Croatia, Romania, Ukraine, and Turkey. Subsequent sequencing with phylogenetic analysis demonstrated that all the isolates were closest to viruses of the Qinghai group, similar to the viruses from Novosibirsk Oblast (Onishchenko et al., 2006; Onishchenko et al., 2006; Onishchenko et al., 2007).

Later, in the winter of 2005-2006, epizootics caused by similar influenza virus strains were observed in Crimea and Ukraine. As it has been shown later by molecular biological investigations, the virus in Ukraine was practically the same as it was in Siberia in the summer and fall of 2005 (see Onishchenko et al., 2007). In February 2006, mortality of wild birds was recorded in Azerbaijan; however, sick and dead birds were secretly picked up and eaten by local residents. As a result, eight human cases with five lethal outcomes were recorded there. The infection was most likely caused when people inhaled aerosolized fecal matter from sick birds, which contained the virus, as they were plucking and cutting bird carcasses.

In March 2006, cases of avian influenza were recorded in both wild and domesticated birds in open-type individual and collective farms in Dagestan and Krasnodar Krai, where more than 1 million birds were killed to stop the spread of the epizootic. Influenza outbreaks among wild birds were recorded in March in Georgia and in western Kazakhstan (the city of Aktau). In April-May, new epizootics were recorded in China, in the Qinghai Lake region and Tibet Province. In May, outbreaks were recorded in western and central Mongolia on Uvs-Nuur Lake. This lake borders Russia; therefore, the outbreaks took place in Russia as well, on the northern coast of the lake. In addition, outbreaks among wild birds and, in some cases, among domesticated birds were recorded in Novosibirsk and Omsk oblasts and Altai Krai in May-June. Following these outbreaks, local offices of Rosselkhoznadzor immediately banned transportation of poultry products to other regions of Russia and abroad. Note that in the spring of 2006, almost all the stock on individual poultry farms in western Siberia was inoculated with inactivated vaccine based on the H5N1-subtype influenza virus, and an epizootic in Novosibirsk Oblast was recorded only in the village of Reshety, Dovolnoye Region, where the inhabitants refused to vaccinate their domestic fowl. Thus, their unintentional experiment demonstrated that the veterinary vaccine used was actually effective (personal communication of Rosselkhoznadzor official).

Later, in June 2006, epizootics were also recorded in Odessa, Kherson, and Sumy oblasts (Ukraine). During the summer, outbreaks occurred in several oblasts of southern Russia, mainly among wild birds but also sometimes on individual farms. All the above data about avian influenza outbreaks in Russia and in neighboring countries have been extracted from specialized Russian Internet sites (*www.rospotrebnadzor.ru* and *fsvps.ru/fsvps/links/structureLinks.html?_language=ru*), nonspecialized Russian Web sites (*www.regnum.ru* and *www.rbc.ru*), and the Web site of the World Organization for Animal Health (*www.oie.int*).

During the spring and summer of 2006, disease outbreaks caused by the H5N1-subtype avian influenza virus were recorded in Ukraine (Odessa, Kherson, and Sumy oblasts). That summer, additional outbreaks were noted in European Russia (Kabardino-Balkaria; Chechen Republic; the republics of Adygeya, Dagestan, and Kalmykia; Stavropol and Krasnodar krais; and Astrakhan, Volgograd, and Rostov oblasts, where mortality was observed both among wild and domesticated birds on individual farms).

It is evident from the phylogenetic tree that we constructed (see Figure 10-3) that the H5 virus strains isolated in 2006 were somewhat different from the strains of 2005 (Lipatov et al., 2007). In addition, the isolates recovered in western Sibe-

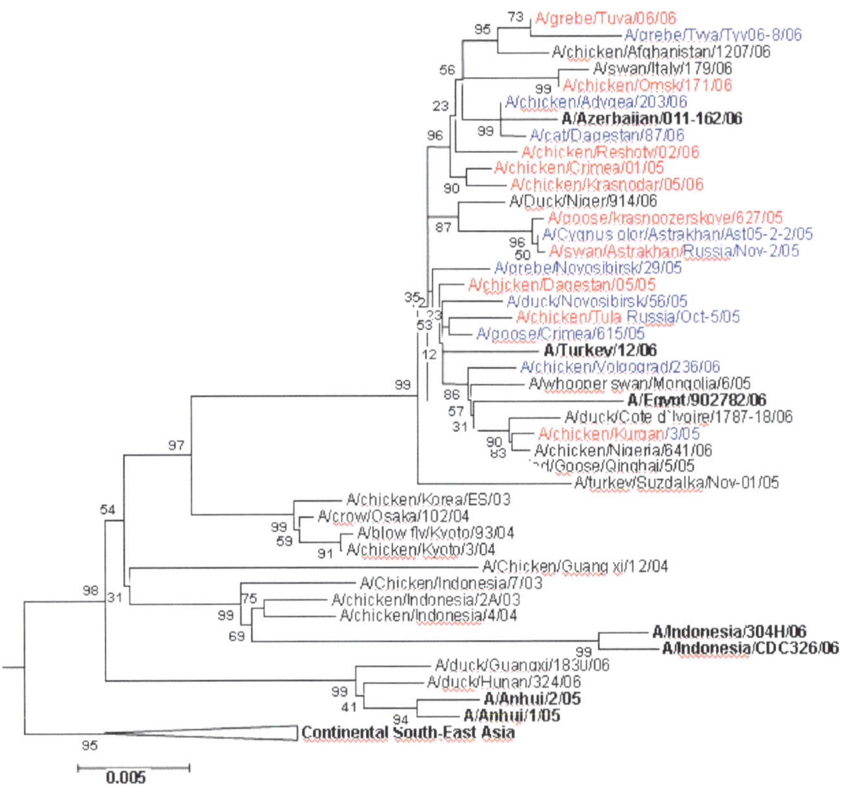

FIGURE 10-3 Phylogenetic tree constructed based on sequences of the HA gene of H5N1-subtype avian influenza virus isolates recovered during 2005-2006 compared with data from other centers.

ria in 2005 were heterogeneous and differed not only in their nucleotide sequences but also in their biological properties. For example, the strains isolated in the village of Krasnoozerskoye and in Dovolnoye Region in Novosibirsk Oblast—only 200 kilometers apart—differ fundamentally in their pathogenicity for mice: 1,000-fold in the infection dose, 10,000-fold in LD_{50} (dose lethal to 50 percent of subjects), and in the virus titers in the lungs, brain, and kidneys of the infected mice (see Table 10-1). Thus, strains pathogenic not only for birds but also for mammals were already circulating in Novosibirsk Oblast in the summer of 2005. Consequently, an avian influenza virus strain pathogenic for humans probably could have been circulating in Siberia at that time; perhaps the population of western Siberia was just lucky to avoid human influenza cases caused by the H5-subtype influenza virus. Note that all severe respiratory disease cases at least in Novosibirsk Oblast and in neighboring regions of western Siberia in 2005-2006 were thoroughly monitored. All samples from such human cases were examined for markers of H5-subtype influenza virus both at local laboratories and, in case of an even slightly positive result, at Vector Center; however, no human cases of H5 avian influenza were detected (Evseenko et al., 2006).

In 2006 the following antiepidemic measures were introduced in western Siberia and all of Russia to control epizootics of influenza virus in birds and to prevent outbreaks of this disease in humans:

- Information was provided about the need to avoid close contact with sick and dead birds (the public, especially in villages, was alerted not only through the media but also with leaflets).
- Acute respiratory disease cases among the rural population were thoroughly monitored, including analysis of suspected cases by real-time polymerase chain reaction (PCR) at Vector and other laboratories in the region. The PCR test was practically identical to the World Health Organization (WHO) test. The Russian PCR test kit was produced by InterLabService, Inc. in Moscow.
- The rural population and high-risk population cohorts in regions with recorded epizootics caused by H5N1 avian influenza virus in birds were vaccinated against seasonal influenza. This vaccination was recommended by the World Health Organization and was conducted in the spring and fall using Russian-manufactured vaccine.
- Domesticated birds on both individual and commercial farms where a mass mortality was recorded and the pathogenic influenza virus was found were exterminated by specially trained teams from the Ministry for Emergency Situations and veterinarians. The carcasses were burned with diesel fuel at special waste disposal sites close to the affected villages or poultry plants, and the sites were subsequently decontaminated with bleach according to procedures specified in the special veterinary biosafety manual.
- More stringent regulations were imposed on commercial poultry farms, requiring the establishment of well-regulated sanitary control measures, preven-

TABLE 10-1 Data on Pathogenicity of Various Avian Influenza Virus Strains in Mice (Onishchenko* et al., 2006; Lipatov et al., 2007)

				Organs (in lg of titer)				
Virus strain	$lgEID_{50}$	$lgMID_{50}$	$lgMLD_{50}$	Lungs	Spleen	Brain	Liver	Kidneys
A/Gs/Krasnoozerskoye/627/05	9.2	2.2	2.3	6.1	1.6	5.2	1.6	2.6
A/Tk/Suzdalka/1-12/05	9.3	5.3	6.3	4.1	<1	2.3	<1	<1
A/VN/1204*	9.8	2.3	3.8	6.9	3.4	2.2	<1	<1
A/Ck/Indonesia/05*	9.3	5.3	>7	4.3	<1	<1	<1	<1

tion of any contacts between domesticated and wild birds, thermal treatment of feed, and so forth.

- Spring and autumn hunting of wild fowl was prohibited or limited, and all hunters were notified of the need to incinerate the intestines and feathers of any fowl taken.
- The most stringent limitations or prohibitions were placed on transportation of live domesticated birds and their meat from region to region.
- Domesticated birds were vaccinated in all areas where avian influenza epizootics were recorded in 2005. The village of Reshety, where the inhabitants refused to vaccinate their birds, was the only site in western Siberia that suffered from avian influenza. Already in 2007, all inhabitants of Novosibirsk Oblast agreed to vaccinate their birds.

In 2007, avian influenza outbreaks were recorded only in the European part of Russia:

- The Republic of Adygeya in Krasnodar Krai: There were several outbreaks among domesticated birds on individual farms in January-February, with the recovered strains shown to be closely related to strains isolated in Azerbaijan and Turkey in the fall of 2006.
- Nine regions of Moscow Oblast in February (in domestic fowl kept in yards): Several suspicious human cases were recorded; however, other causes of acute respiratory diseases were found. The outbreak in Moscow Oblast was most likely caused by poultry illegally transported from Krasnodar, as was demonstrated by examining the nucleotide sequences of recovered avian influenza virus isolates.
- Krasnodar Krai in September (in domesticated birds on a private farm)
- Rostov Oblast in December (in domesticated birds on individual farms)

Note that stringent sanitary measures in combination with the vaccination of domesticated birds and other measures considerably decreased the number and scale of epizootics recorded in European Russia in 2007 and allowed epizootics in Asian Russia in 2007 to be avoided entirely.

It should also be highlighted that the situation regarding this disease in wild and domestic fowl in China in 2007 was rather unusual: No epizootics were recorded in the Qinghai Lake region. Presumably that is why migratory birds flying north in the spring to Russia and Kazakhstan carried no influenza viruses. On the other hand, no epizootics occurred in the Chinese provinces where human cases were recorded, which is quite unusual. There were reports in the media about the mass vaccination of domestic birds in China in the spring of 2007, but authors have no data about the scale of the vaccination effort or the vaccine composition and type.

Thus, humankind has sufficient measures for controlling epizootics caused by the H5N1 avian influenza virus. Mortality of domesticated birds on individual farms can be minimized and on commercial closed-type poultry farms can be completely excluded in countries where all antiepidemic measures are carried out, the public is kept informed about ways of minimizing the risk of domestic fowl infection, and birds are vaccinated.

However, it should be kept in mind that avian influenza is not the only emerging disease and that migratory birds are only one potential source for the appearance and spread of emerging infections. Other possible sources are discussed in the following sections.

HUMAN COLONIZATION OF NEW TERRITORIES INHABITED BY ANIMALS OR INSECTS PREVIOUSLY UNKNOWN TO HUMANS (TICK-BORNE ENCEPHALITIS IN 1937-1940 IN THE RUSSIAN FAR EAST)

When people enter new and previously unexplored territories, pathogens can be transferred from animals or insects to humans. This happened in the late 1930s in the Russian Far East during construction of the Khabarovsk–Komsomolsk-na-Amure railroad. During this project, tens of thousands of people worked in the taiga, where virtually no people had been present before. During the first year of construction, mass human cases of encephalitis were recorded. To clarify the underlying reasons, several expeditions of researchers and experts headed by the outstanding Russian scientists L. A. Zilber, E. V. Shibladze, E. Pavlovsky, A. A. Smorodintsev, and M. P. Chumakov were sent to the area. The expeditions discovered that ticks were the vector for the pathogen in question. Measures for controlling ticks and avoiding tick bites were immediately implemented. Special clothing was designed for workers, and the practice of regular mutual tick examinations every 3-4 hours was introduced. Consequently, the morbidity rate was considerably reduced. Later, a vaccine against this disease was developed. This disease is today controllable by vaccination, and only the insufficient level of vaccination among the population is the reason for tick-borne encephalitis morbidity in Russia (up to 7,000 cases annually). However, it should be noted that tick-borne encephalitis virus continues to spread not only in Russia but also in other countries in northern and central Europe and Asia, including Germany, Austria, Switzerland, the Czech Republic, Kazakhstan, and others. Vaccination against tick-borne encephalitis is therefore becoming increasingly widespread in these countries.

INDUSTRIAL BREEDING OF RARE ANIMAL SPECIES (PALM CIVET AND SARS CORONAVIRUS)

Today it is well known that the commercial breeding of rare palm civets for their meat was the source of severe acute respiratory syndrome caused by SARS coronavirus in China. The Chinese recently started eating palm civet meat and breeding the animals, and civets frequently carry the coronavirus. Researchers have discovered that a random deletion of an insignificant portion of the gene encoding a key protein (less that 0.1 percent of the genome) and several nucleotide substitutions made this coronavirus infectious for humans. When consumption of civet meat and commercial breeding of the animals were halted, human contact with these animals also stopped, as did the epidemics caused by the coronavirus in question.

INTRODUCTION OF NEW ANIMALS TO NEW TERRITORIES

Several animal species—muskrat, American mink, and nutria—were imported to Russia in the 1930s, initially for captive breeding. These species later successfully acclimatized in the wild. Cases of Omsk hemorrhagic fever appeared in the area inhabited by muskrat 20 years after this species appeared there. Sequencing of the genome of Omsk hemorrhagic fever virus demonstrated that it was very closely related to the tick-borne encephalitis virus. During the past 30 years, cases of Omsk hemorrhagic fever were recorded only among muskrat hunters. Presumably, this means that muskrat had become the natural host of this virus.

As for American mink and nutria, we do not yet know the particular pathogens that can be transmitted from these animals to humans; however, it is quite possible that scientists may discover either new or changed pathogens that came to Eurasia from the American continent with mink or nutria or that have changed during passaging in these species.

CLIMATE CHANGE (WEST NILE AND JAPANESE ENCEPHALITIS VIRUSES IN SIBERIA)

Global warming and resulting climate change creates conditions appropriate for reproduction of more southern insect species on territories with previously severe climate, and these insect species that are new to particular areas appear able to transmit diseases that have not been transmitted by the insects that have long inhabited these areas. As just one example, West Nile virus is now detectable not only in the European part of Russia but also in western Siberia and the Russian Far East; moreover, it is found not only in birds and mosquitoes but also in human encephalitis cases. Most likely, this represents a more global process than seemed the case 2 or 3 years ago, as a considerable level of antibodies to this virus was detected in the human population in 2007, amounting to 20 percent of the population in certain regions of western Siberia. In addition, isolated cases of Japanese encephalitis are being recorded in the Russian Far East (earlier, this disease was very rare in Siberia). All these facts suggest that climate change as a result of global warming not only can cause thawing of permafrost and ice but can also lead to the emergence of tropical diseases in new areas due to propagation of insect species that are vectors of these diseases.

UNINTENTIONAL CREATION OF NEW CONDITIONS FOR PROPAGATION OF ANIMALS AND INSECTS

It is known that dog and wolf populations in trash dumps increase rapidly if not controlled, thereby allowing for reproduction of dangerous diseases such as rabies. Unfortunately, this situation has occurred recently in several cities of western Siberia, and only intensive control of stray dogs has reduced the number of infected animals in the neighborhoods of these cities.

As for other analogous processes, experts noted long ago that mosquitoes reproduce very intensively in used tire dumps because of favorable conditions created inside the tires. This situation can also be threatening for Siberia, as crowds of mosquitoes appearing in tire dumps in combination with the animals inhabiting the same sites can form reservoirs for a multitude of infections, including malaria. Fortunately, malaria is still not endemic in Russia, but if further climate change occurs, it may become endemic, which would have dramatic consequences given the huge territory and many lakes, ponds, and marshes in Siberia and in Russia as a whole.

ADOPTION OF NEW TECHNOLOGIES

Adoption of household technologies new to a particular region can also be a reason for the emergence of new infections. In particular, it was known in Russia that Legionnaire's disease cases were recorded abroad; however, no cases were found in this country until recently. On the other hand, humidifying air conditioners were virtually absent in Russia until the 1990s; therefore, the specific conditions for cultivation and reproduction of the corresponding bacteria were also absent. Today, such air conditioners are present not only in offices but also in apartments, so cases of Legionnaire's disease have consequently appeared. In western Siberia, such cases were recorded in Biisk (Altai Krai) 3 years ago. Several dozen Legionnaire's disease cases were recorded in Yekaterinburg and neighboring cities in 2007. Some were associated with air conditioners and some with stagnant warm tap water that was also polluted, conditions that enhanced reproduction of legionellosis bacteria and subsequent human infection.

CONCLUSIONS

Reproducing intensively and colonizing new territories, humankind itself creates new possibilities for the reproduction, spread, and variation of infectious pathogens. Correspondingly, it is necessary to take into account all the possible reasons that bring about new infections in order to prevent their emergence or minimize their consequences. In particular, monitoring acute zoonotic and potentially zooanthroponotic infections in wild animals and birds in areas close to their habitats and migratory pathways in Russia and the other countries of the Commonwealth of Independent States (CIS) is most useful for preventing the spread of emerging infections. This is also very important for European countries, as migratory birds during one season transfer the pathogens reproducing in them over vast territories. One of the most important rest stops and nesting grounds for birds migrating to Eurasia is located in the southern part of western Siberia (Chany Lake and other lakes of Altai Krai and Omsk and Novosibirsk oblasts). Therefore, it is most important to monitor and study the pathogens of various infections in migratory birds and wild animals in these particular regions, as they will appear in these regions somewhat earlier than they will become dangerous for people. Further strengthening Russian research potential in this field is vital for providing early alerts of new emerging infections for Russia, other CIS countries, European and Asian countries, and even the United States and Canada, as the so-called Palearctic migratory flyway goes from Siberia to Alaska, so migratory birds can potentially transfer pathogens all over the American continent. Therefore, joint research on infectious agents, especially zoonotics, in Russia and the United States will assist both countries in countering new threats of emerging infections. We are well aware that only one emerging infection—avian influenza—has claimed more than 160 lives during the past 6 years, whereas bioterrorism is to blame for only six deaths during that time. This means that Nature is still the world's chief bioterrorist. An increase in our joint potential in the control of emerging and yet unpreventable infections will contribute to public health in our nations and in neighboring countries and provide us with more options for combating any kind of bioterrorism, be it deliberate or generated by nature.

REFERENCES

Evseenko, V. A., A. V. Zaikovskaya, V. A. Ternovoi, A. G. Durimanov, S. I. Zolotykh, Y. N. Rassadkin, A. S. Lipatov, R. G. Webster, A. M. Shestopalov, S. V. Netesov, I. G. Drozdov, and G. G. Onishchenko. 2007. Diversity of highly pathogenic avian influenza H5N1 viruses that caused epizootic in western Siberia in 2005. Doklady Biological Sciences 414:226-230.

Ilyinskikh, E. N., I. N. Ilyinskikh, and A. V. Lepekhin. 2008. The first cases of West Nile fever in Tomsk region. Materials of the scientific and practical conference "Actual problems of tick borne infections." Medicine in Kuzbass 6:75-76 (in Russian).

Kononova, Yu. V., A. G. Mirzaeva, Yu. L. Smirnova, E. V. Protopopova, T. A. Dupal, V. A. Ternovoi, Yu. A. Yurchenko, A. M. Shestopalov, and V. B. Loktev. Species composition of mosquitoes (Diptera, Culicidae) and possibility of the West Nile virus natural foci formation in the south of Western Siberia. Parasitology 41(6):459-470 (in Russian).

Lipatov, A. S., V. A. Evseenko, H. L. Yen, A. V. Zaikovskaya, A. G. Durimanov, S. I. Zolotykh, S. V. Netesov, I. G. Drozdov, G. G. Onishchenko, R. G. Webster, and A. M. Shestopalov. 2007. Influenza (H5N1) viruses in poultry, Russian Federation, 2005-2006. Emerging Infectious Diseases 13(4):539-546.

L'vov, D. K., M. Yu. Shchelkanov, P. G. Deriabin, T. V. Grebennikova, A. G. Prilipov, E. A. Nepoklonov, G. G. Onishchenko, N. A. Vlasov, T. I. Aliper, A. D. Zaberezhny, D. E. Kireev, O. P. Krasheninnikov, S. T. Kiriukhin, E. I. Burtseva, and A. N. Slepushkin. 2006. Isolation of influenza A/H5N1 virus strains from poultry and wild birds in West Siberia during epizooty (July 2005) and their depositing to the state collection of viruses (August 8, 2005). Advances in Virology 51(1):11-4 (in Russian).

Onishchenko, G. G., S. P. Bereznov, A. M. Shestopalov, A. Yu. Alekseev, V. A. Ternovoi, A. B. Khaitovich, M. T. Krovyakova, S. V. Netesov, and I. G. Drozdov. 2007. Molecular biologic analysis of avian influenza virus isolates which caused epizootics on the south of West Siberia and in Crimea. Zh Microbiology, Epidemiology, and Immunobiology 5:28-32 (in Russian).

Onishchenko, G. G., A. M. Shestopalov, V. A. Ternovoi, V. A. Evseenko, A. G. Durimanov, Yu. N. Rassadkin, Yu. V. Razumova, A. V. Zaikovskaya, S. I. Zolotykh, S. V. Netesov, and L. S. Sandakhchiev. 2006. Highly pathogenic influenza virus H5N1 found in western Siberia is genetically related to viruses that circulated in Southeast Asia in 2003-2005. Doklady Biological Sciences 406:63-65.

Onishchenko, G. G., A. M. Shestopalov, V. A. Ternovoi, V. A. Evseenko, A. G. Durimanov, Yu. N. Rassadkin, A. V. Zaikovskaya, S. I. Zolotykh, A. K. Yurlov, V. N. Mikheev, S. V. Netesov, and I. G. Drozdov. 2006. Study of highly pathogenic H5N1 influenza virus isolated from sick and dead birds in Western Siberia. Zh Microbiology, Epidemiology, and Immunobiology 5:47-54 (in Russian). Erratum in Zh Microbiology, Epidemiology, and Immunobiology 7:128.

Shestopalov, A. M., A. G. Durimanov, V. A. Evseenko, V. A. Ternovoi, Yu. N. Rassadkin, Yu. V. Razumova, A. V. Zaikovskaya, S. I. Zolotykh, and S. V. Netesov. 2006. H5N1 influenza virus, domestic birds, western Siberia, Russia. Emerging Infectious Diseases 12(7):1167-1169.

11

A Note on the Interfacial Vulnerabilities of Transportation Systems

George Bugliarello, Polytechnic University and National Academy of Engineering

Some of the most insidious vulnerabilities of a system often are encountered at the interfaces within the system and between the system and other systems. In transportation systems these vulnerabilities are aggravated by a situation in which not all their components and interfaces are being paid equal attention. For instance, there are today no agent-based models that incorporate realistically the human component of the system and can predict the emergency response of individuals.

The components of a transportation system range from individuals in the system—a biological component—to its social components, such as its organization and practices, to its machines, that is, in the broadest sense, its engineered artifacts, as described in Table 11-1. Thus, transportation systems are complex bio-socio-machines, or, for short, *biosoma* systems, in which their biological, social, and machine components are indissolubly interconnected (see Figure 11-1) (Bugliarello, 2000; Bugliarello, 2003). Also, *biosomic*[1] is the range of modalities of potential attacks on such systems and so are many of their interfacial vulnerabilities and their resilience.

TABLE 11-1 Biosoma Components of a Transportation System

Bio	So	Ma
Individual operators	Organizations operating systems	Vehicles
Managers	Security forces	Platforms (roads, rails, pipelines, etc.)
Security personnel	Government Local Regional National	Access facilities (station, harbor, airport, etc.)
Individual users	International compacts	Power and fuel supply
Working animals	Community	Sensors, command-and-control devices

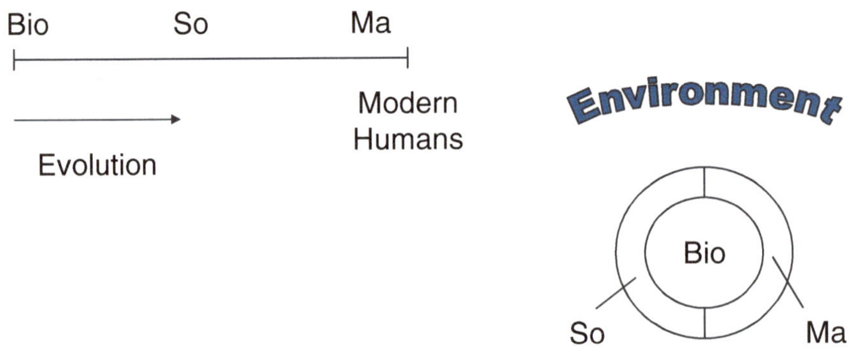

FIGURE 11-1 The biosoma.

ATTACKS AND VULNERABILITIES

Attacks on transportation systems can occur from individuals acting independently, such as hackers or some suicide bombers, or can be organized by social entities, such as terrorist groups of various degrees of sophistication, for example, those responsible for the September 11, 2001, attack in New York City and the attacks on trains in Madrid, London, and Mumbai. The attacks can be carried out by different kinds of machines, from conventional explosives to poisonous chemicals, nuclear or radiological devices, vehicles, airplanes, and generators of electromagnetic pulses. In general, the tighter the interconnections among the biosoma components of an attack, the more effective the attack is likely to be.

VULNERABILITIES WITHIN A TRANSPORTATION SYSTEM

The interfacial vulnerabilities within a transportation system are those at the interfaces between components of the biosoma—between individuals and organizations, between individuals and machines, or among all three biosoma components. The individual-organization vulnerabilities could be due, for example, to vulnerabilities in the relation between a conductor and the management of the system, such as failures to communicate with each other, to alert each other, to supervise, or to accept supervision. The vulnerabilities often occur because of psychological conditions, such as that of a disaffected individual or organization; of an individual fearing for his or her family; language barriers that may lead to misinterpretations; or physiological conditions, such as the health of the individual or, for that matter, the gestalt of an organization that is dysfunctional or not performing at its best. Examples of vulnerabilities at the individual-machine interface are inadequate operational knowledge of an individual operating a machine, machine failures, or sabotage of a machine by an ill-intentioned individual.

An example involving the interfaces among all three biosoma components within the system is failures of the traffic light command-and-control system involving incident commanders, the organization that operates the traffic lights, and the traffic lights themselves. Critical and choke points of a transportation system, such as locks, canals, traffic light control centers, and other command-and-control centers, are particularly vulnerable to biosomic attacks that may involve individual sabotage, organized attacks, and the use of machines to destroy critical components of the system. One of the recent examples of vulnerabilities involving all biosomic components and the disconnects at their interfaces was the response to Hurricane Katrina in 2005 in New Orleans. There were failures of individuals to take action—both decision makers in positions of leadership and victims unwilling to evacuate; failures of organizations such as the emergency response systems; and failures of machines, such as levees, houses, and bridges that collapsed. The most egregious failure, however, was the lack of coordination across the multiple interfaces among these components, for example, coordination among evacuation organizations, the viability of bridges, and the availability of vehicles. This exacerbated the individual failures of the components.

KEY SYSTEMS INTERDEPENDENCIES

The vulnerabilities of a transportation system are also affected by those of other systems with which it interfaces. The multiple interfaces of transportation systems are summarized in Box 11-1. The telecommunications system enables a transportation system to communicate within itself and with other systems; the energy system provides the power and fuel the transportation systems require; the financial systems are involved in collection of revenues, payment of salaries,

Box 11-1
Transportation Systems' Interfaces

- Interfaces Within a System
  - Individuals (BIO); organizations, regulations, practices, etc. (SO); and machines: vehicles, infrastructure (roads, docks, power, communications, sensors, etc.) (MA)
- Interfaces With Other Systems
  - Other Transportation Systems (land, water, and air)
  - Other Systems (power, telecommunications, security, finance, international trade agreements, etc.)
- Jurisdictional Interfaces
  - Municipal
  - Regional
  - National
  - International
- Private-Public Interfaces

financing of construction and repair, and so forth; and security systems help provide protection. Furthermore, transportation systems may be affected by the systems of international trade agreements, which affect, for instance, the inspections at the source or point of shipment of merchandise coming from other countries.

Transportation systems, in turn, have an impact on other systems—on their human resources and on their flow of needed supplies. The impacts on human resources may stem from the possible inability in an emergency for a transportation system to convey the first responders to their assigned locations, as well as to convey personnel necessary for the operation of business, industry, education, health care, and security systems. The impact on supplies occurs when the transportation system cannot deliver food, materials required by energy systems and industry, and other supplies needed by businesses and service systems.

These interfacial interactions may involve all three components of the biosoma—the individuals (for example, the conductors and the managers), the organizations (such as the transportation system itself, the telecommunications systems, or the system of first responders), and the machines (such as trains, pipelines, and information systems). Figure 11-2 exemplifies some of these intra- and intersystem interactions.

Some important vulnerabilities of transportation systems occur at intermodal interconnections, for example, within or among road systems, rapid transit systems, railroad systems, water transport systems, and air transport. Vulnerabilities can arise, for example, when there are disconnects between freight in harbors

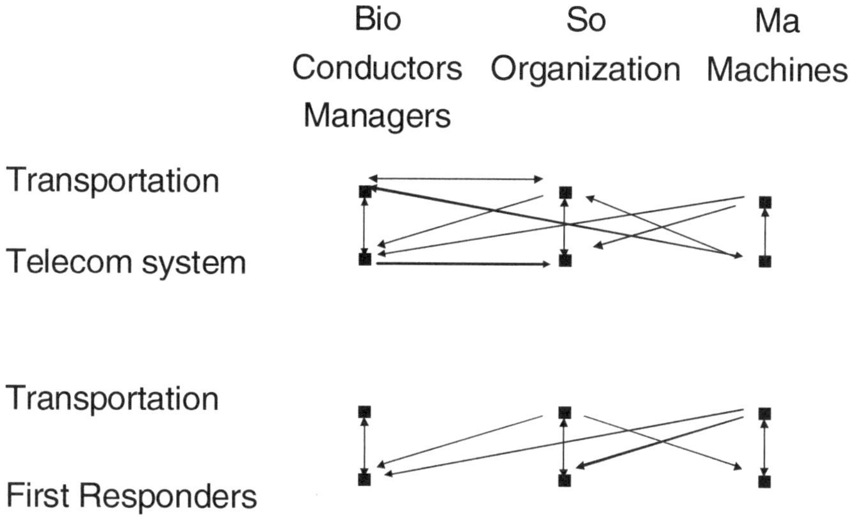

FIGURE 11-2 Examples of patterns of system interactions.

and freight at freight terminals, as containers are landed and then transferred to rail or truck. In addition to the vulnerabilities at each of these interconnections, there are issues of coordination of the overall transportation system and of feasibility of substitutions if one component of the system, for example, a railroad segment, fails.

Transportation systems are particularly vulnerable at their people access points: at stations, such as ticket offices, waiting rooms, and platforms that often gather crowds; at bus and street car stops; at taxi stands; and at airports, where long lines typically precede the access to security.

Identifying the vulnerabilities of the points of contact among the *organizational* components of the transportation system and its interfaces with other organizations and jurisdictions is particularly important. There is a need to consider the potential vulnerabilities of individuals involved in the organizational interfaces (vulnerabilities that might arise from their state of health or mind, or language barriers that may lead to misinterpretations), as well as the potential organizational vulnerabilities that may stem from different practices or views in different organizations of what is correct practice. These interorganizational vulnerabilities are exemplified by the recent case of a traveler diagnosed with a highly infective case of tuberculosis who was cleared to cross the border from Canada to the United States in spite of an all-points border alert. However, in all these cases judgment and the avoidance of paranoia are also needed.

There is also a need to consider the vulnerabilities arising from the interfaces among different jurisdictions. A transportation system usually spans different jurisdictions, as is seen in railroads, pipelines, or international airlines operating across several national borderlines or of commuter railroads crossing boundaries between municipal jurisdictions. Jurisdictional problems have obvious implications for security, for example, when an incident calls for the involvement of multiple jurisdictions. A frequent issue is the lack of coordination, when jurisdictions that should have intervened immediately defer to other jurisdictions, for example, to local entities, that may not have the capacity to take preventive measures or to mitigate the consequences of the disaster.

Some vulnerabilities of transportation systems stem from economics and financial issues. One such set of issues involves the trade-off between security and efficiency; that is, whether the system should become more centralized or less centralized, whether several systems, such as pipelines and telecommunications lines, should be colocated, whether the organizational structure of the systems may or may not be geared to optimally deal with a major disaster, and whether, as is increasingly the case, the system relies on using the public Internet with its cybervulnerabilities rather than dedicated communications, which are less efficient but more secure. A second set of issues stems from the interfaces between public and private components. In the United States, the ownership of transportation systems, as that of most infrastructural systems, is to a very large extent private, hence, introducing vulnerabilities in their interactions with public systems. These trends are exacerbated by the growing internationalization of transportation systems, which further complicates the assessment of vulnerabilities and the defense against them. An examination of the critical issues in transportation in the United States identified by the Transportation Research Board of the National Academies—several of them in the economic and financial domains—would show that they require in most cases an interlaced biosomic approach, with its associated interfacial vulnerabilities (TRB, 2006).

RESILIENCE, PREVENTION, AND MITIGATION

When a disaster occurs, whether anthropogenic or natural, in which the vulnerabilities of a system reduce its functionality, the system's resilience, that is, the ability to recover its functionality fully or in part, within a reasonable time, is of paramount importance (see Figure 11-3). It is determined by the resilience of its individual components and by the effectiveness of their biosomic interactions.

The first step in prevention is to identify all the interfaces of the system, both internal and external, assess for each the probability that an incident may happen at each interface and determine its consequences—the extent to which the interface might be damaged or totally taken out of commission. That assessment of probabilities and consequences needs to focus not only on the most probable events but also on those of lower probability that may have much greater consequences. An important part of the assessment is the robustness of the interfaces, in order to guide decisions about preventive hardening, and about possible bypass-

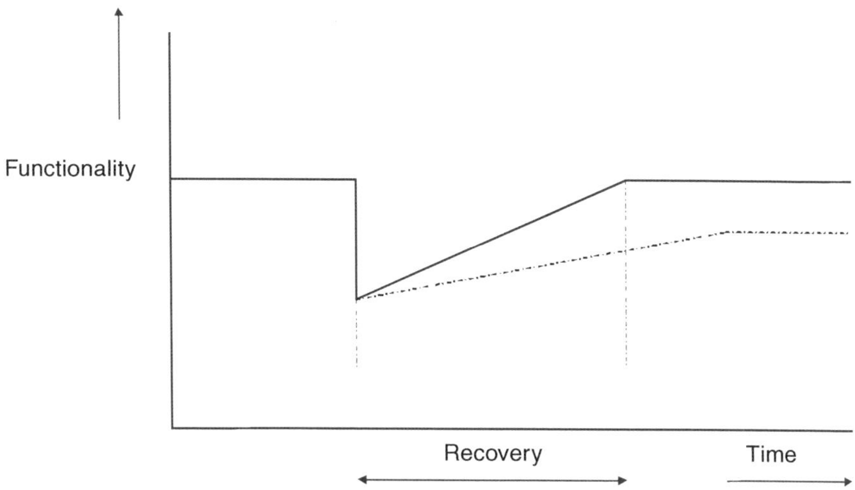

FIGURE 11-3 Resilience.

ing of the interfaces through connections with other components of the system or with other systems, for example, a different mode of transportation.

VULNERABILITY MAPS

Idealized computerized maps of a system can be useful in describing in a synoptic way different interfacial vulnerabilities, both within the system and where it interfaces with other systems (such as access points—stations and loading docks; supply points, where the system interfaces with the power and fuel supply systems; intermodal transfer points; maintenance and repair facilities) and where it interfaces with the environment surrounding the system (such as phone lines, viaducts, and so forth), which may be points of joint physical vulnerabilities with other systems (see Figure 11-4). The vulnerability of each of these interfacial points needs to be assessed for all the biosoma components of the interfacing systems. For instance, for a station, assessment should include the vulnerabilities of passengers to individual terrorists, the effectiveness of the station management organization, the robustness of the station's structures, the vulnerabilities at the interfaces between the station, the vehicles in it and outside of it, and the communications system that will be informing the conductor of the vehicles.

An assessment of the risks associated with a potential disruption at each of the vulnerability points on the map can be incorporated in a multidimensional representation of the risks throughout the system. This can make it possible to prioritize biosomic preventive measures and mitigating actions, including, if nec-

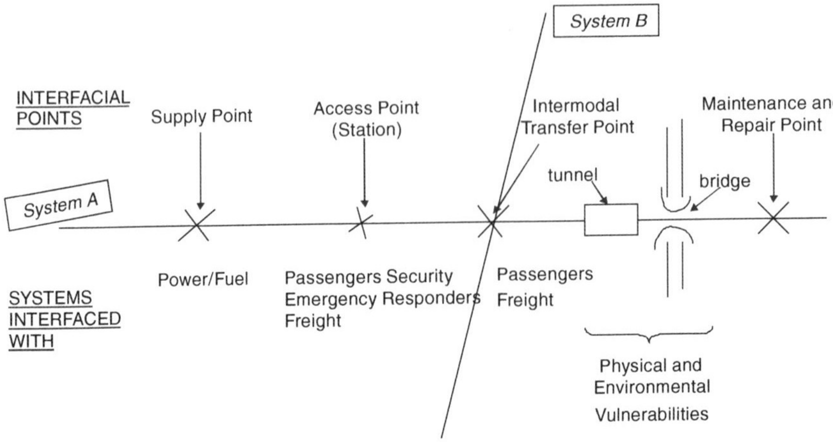

FIGURE 11-4 A vulnerability map of a transportation system.

essary, reconfiguration of the system to bypass disrupted interfaces. However, the process is not straightforward, as the assessment of risk depends on the realism of possible scenarios of disruption, on an estimate of their probabilities and of their consequences—a difficult problem for events for which there is no precedent. The assessments and the actions based on them can become further complicated if multiple simultaneous or sequential disruptions occur.

CONCLUSIONS

1. The interfacial vulnerabilities of a transportation system are multiple and complex.

2. A view of transportation systems as bio-social-machine (biosoma) systems helps identify systematically their interfacial vulnerabilities, both those internal to the systems and those in the systems' interaction with other systems.

3. The effectiveness of terrorist attacks on a transportation system also depends on the coordination of their biosoma components and so does the system's resilience.

4. Research is needed to go beyond empirical approaches and to develop a new body of knowledge that would help identify the vulnerability of interfaces, assess the probability and potential consequences of events at interfaces, study potential cascade effects, and develop strategies and technologies to bypass compromised interfacial points. All this requires more systematic and rigorous research to better understand the vulnerabilities of transportation systems and other infrastructural systems and to develop possible remedies. For example, a better understanding of the interactions of human and nonhuman biosoma components

of transportation systems could benefit from the emerging interdisciplinary field of social network analysis (Heyman, 2006).

ACKNOWLEDGMENTS

This research is supported in part by Sloan Foundation Grant number 2002-10-12 to the Urban Security Initiative at Polytechnic University.

NOTE

1. Pertaining to the biosoma.

REFERENCES

Bugliarello, G. 2000. The biosoma: The synthesis of biology, machines and society. Bulletin of Science, Technology & Society 20(6):452-464.

Bugliarello, G. 2003. The Biosoma: Reflections on the Synthesis of Biology, Society and Machines. Brooklyn, N.Y.: Polytechnic University.

Heyman, K. 2006. Making connections. Science 313:604-606.

Transportation Research Board (TRB) of the National Academies. 2006. Critical Issues in Transportation. Washington, D.C.: The National Academies Press. Available online at *onlinepubs.trb.org/Onlinepubs/general/CriticalIssues06.pdf*. Accessed April 24, 2008.

12

Transportation Planning for Evacuations

John C. Falcocchio, Polytechnic University

Evacuation planning is a key component of security planning. Effective evacuations require reliable transportation systems capable of moving people out of the danger zone and into safety in a timely manner. This paper addresses the transportation issues to be considered in evacuation planning and highlights the challenges that must be met to develop effective evacuation plans.

TRADITIONAL TRANSPORTATION SYSTEM PLANNING AND DESIGN PRACTICES

The traditional criteria that guide the *planning* and *design* of transportation systems include mobility, safety, accessibility, cost, environmental issues such as air quality, and so forth.[1] These criteria were developed to meet the social and economic transportation needs of society under pre–September 11, 2001, conditions.

For decades we have focused on keeping transportation systems costs down by promoting efficiency. This goal needs to be reviewed because it does not recognize that in emergency evacuations redundancy in the transportation system is essential to keep it resilient. In light of the new reality with security concerns, we need to promote a new system perspective in planning and financing critical transportation infrastructure.[2]

In many of the nation's major metropolitan areas, roadway redundancy is woefully inadequate and would pose serious threats to large segments of the population if interstate highways or other primary arteries were disabled by a terrorist attack and massive evacuations became necessary. The U.S. Conference of Mayors recently released findings of a survey indicating that U.S. metropolitan areas are not prepared for major emergencies and homeland security.[3]

Traditional transportation management practices tend to focus on commuter mobility and the efficient and reliable movement of freight. Advanced technologies[4] are used to monitor system performance and to provide traveler advisories; transportation engineers and managers are able to operate existing systems more efficiently using real-time information management strategies. The operating environment for these functions consists of predictable travel patterns and expected perturbations created by random incidents. The roles of agencies (that is, departments of transportation, police, and emergency services) are coordinated to respond to recurring daily events, such as incident removal, enforcement, roadway management, coordination, and timing of signals. Professionals and other staff responsible for these functions are typically trained in this environment.

Experiences with the terrorist attacks of September 11, 2001,[5] and Hurricanes Katrina and Rita,[6] however, have demonstrated that the traditional transportation planning, design, and management processes that work well under normal conditions are not adequate in responding to the needs of emergency evacuations. Recent experience in evacuating populations during Katrina and Rita demonstrates a lack of adequate preparation for emergencies and points to the need for improving the planning and design of transportation systems for meeting the transportation needs of the population to be evacuated from dangerous areas to safe areas during an emergency. Recent experience also indicates a need for achieving better coordination between first-responder agencies and between different political jurisdictions. There is little doubt that transportation agencies need to reexamine their mission statements with the objective of making emergency evacuation planning an integral part of their work programs.

TRANSPORTATION SECURITY PLANNING FRAMEWORK

Transportation planning for evacuations is an integrative process of coordinating key functions of transportation agencies, first responders, different jurisdictions, and different levels of government. The objective of evacuation planning is to transport to safety the population affected by a natural or engineered life threatening event.

A key element of evacuation planning is transportation security. Figure 12-1 proposes a framework for analyzing the vulnerability of transportation systems to natural or engineered threats and to assess their potential impacts on the system's performance and damage to life. The risk assessment analysis, together with available financial resources, will guide decision makers in establishing strategies

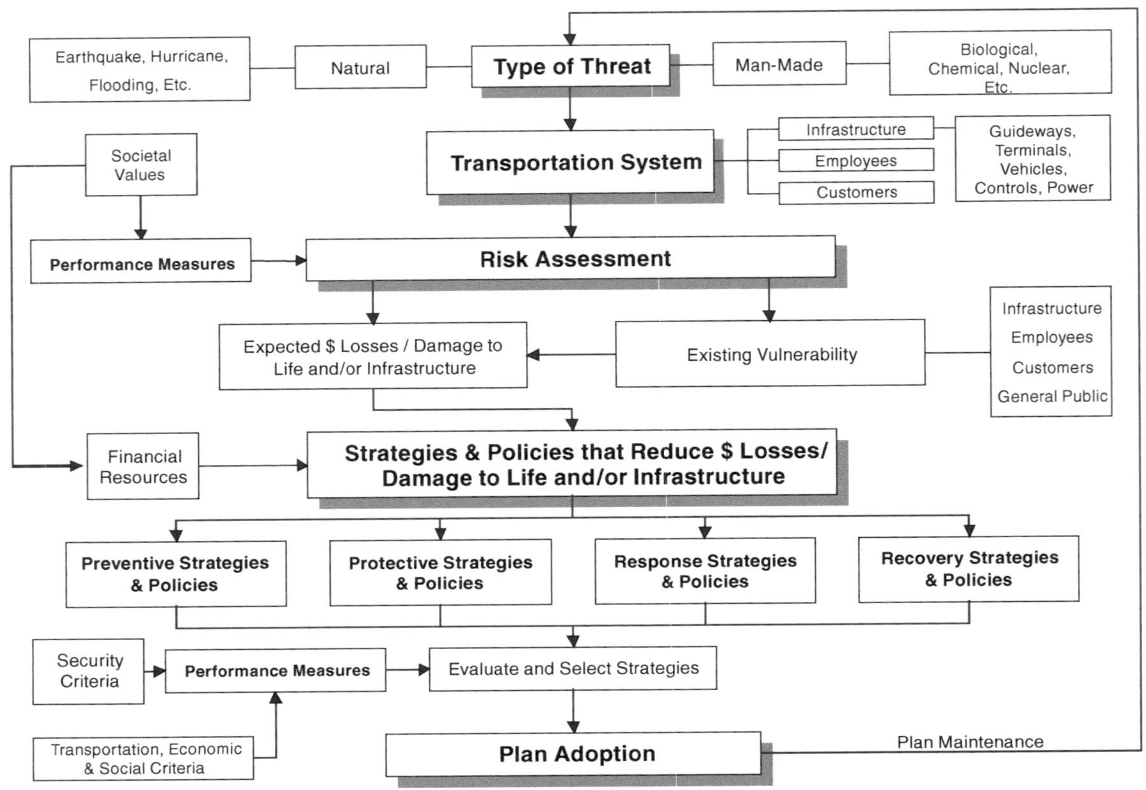

FIGURE 12-1 Transportation security planning framework.

and policies that will reduce damage to life and protect the transportation infrastructure through *preventive* (counteracting potential threats before they occur), *protective* (minimizing the consequences of attacks), *response* (after an attack is under way or has occurred), or *recovery* (bringing the transportation system back to normal) actions.

In evacuation planning, we are concerned with strategies and policies that *protect* the transportation system so that it can perform its basic functions (providing mobility and access) in an effective manner, as well as strategies and policies that enable the development and deployment of operational plans for evacuating people to safety and bringing first responders and their equipment (*response*) to the areas affected by the event.

This paper describes the evacuation planning process in two parts: (1) the plan preparation phase and (2) the evacuation phase. For each phase, key performance criteria will be identified to guide plan making, deployment, and monitoring of the transportation system during the operational phase. Issues in each phase will be addressed to highlight existing practices, and changes to existing practices will be proposed to create better evacuation responses.

PLAN PREPARATION PHASE

The transportation system is a key component of emergency management. A functioning transportation system (providing mobility and access) is fundamental in bringing personnel and equipment to a disaster site and evacuating people from the area. This includes roadways (highways, bridges, tunnels), transit for people without car access, and specialized transportation for those needing assistance. Similarly, evacuation from buildings and activity centers requires having the necessary capacity of exit routes to meet the time constraints of the evacuation.

There are three key elements to be considered in this phase: (1) the configuration of the highway network and its adaptability to respond to changes in management policies during an emergency, (2) the design elements of the highway system that maximize its flexibility and adaptability in meeting emergency conditions, and (3) the organizational preparedness of all agencies in managing the highway system for the movement of first responders and for the evacuation of the population at risk. Examples of each element are indicated below.

Configuration of the Highway Network

The transportation network needs to provide sufficient capacity to serve the demands of evacuation—for the evacuees as well as first responders. This requires establishing evacuation corridors and maintaining lane continuity along major expressways. Existing highway networks need to be modified during emergencies to allow only movements that expedite evacuation, and certain movements that will create bottlenecks in the system should be closed. The system design should allow for maximizing throughput capacity in the evacuation corridor. In this regard, it is essential that different jurisdictions work together to ensure a highway system across jurisdictions with an integrated capacity. Metropolitan planning organizations could play a major role toward this goal and should become involved in security activities.[7]

One potential solution in providing mobility for evacuation needs is to extend the popular highway high-occupancy lanes (HOV) concept into an evacuation special lanes (ESL) network. The popularity of HOV lanes in North America is steadily increasing and policy makers should consider extending the role of HOVs into serving evacuation needs. For example, New York City, Houston, and New Orleans could develop a network of ESLs with highway-to-highway connectivity and central business district coverage to meet disaster management needs. Currently, these cities lack design and operational connectivity, although Houston has made some progress in developing direct ramps to achieve connectivity.[8]

Flexibility in Expressway Operations

Examples of highway-based actions to increase the operational flexibility of the highway system are listed below. The purpose of these improvements is to provide for the special needs of special responder vehicles, access points, navigational needs, lane directional changes to meet travel flow requirements, and information displays. Possible design actions include the following:

- Contraflow lanes options
- Median breaks at crossover points
- Direct ramps for contraflow lanes
- Movable median dividers at critical sections
- Bottleneck bypasses
- Variable message (VMS) at access points
- Driver information at roadside, such as advisory radio frequency for up-to-date traffic condition reports
- Closed-circuit television (CCTV) monitoring
- Lane widths adequate for the movement of large vehicles of first responders
- Roadway sensors and detectors connected to central monitoring system
- Provision of locations for emergency fuel supply along roadway

In a recent study of hurricane evacuation needs (see Figure 12-2) in New York City,[9] it was found that the criteria for locating VMS and CCTV cameras in its expressway system are based on commuter flow patterns and do not reflect the needs of the evacuating population, as demonstrated by the lack of coverage along the routes in the evacuation zones located in the Rockaways area of Queens (see Figures 12-3 and 12-4). This finding points to the need for rethinking the

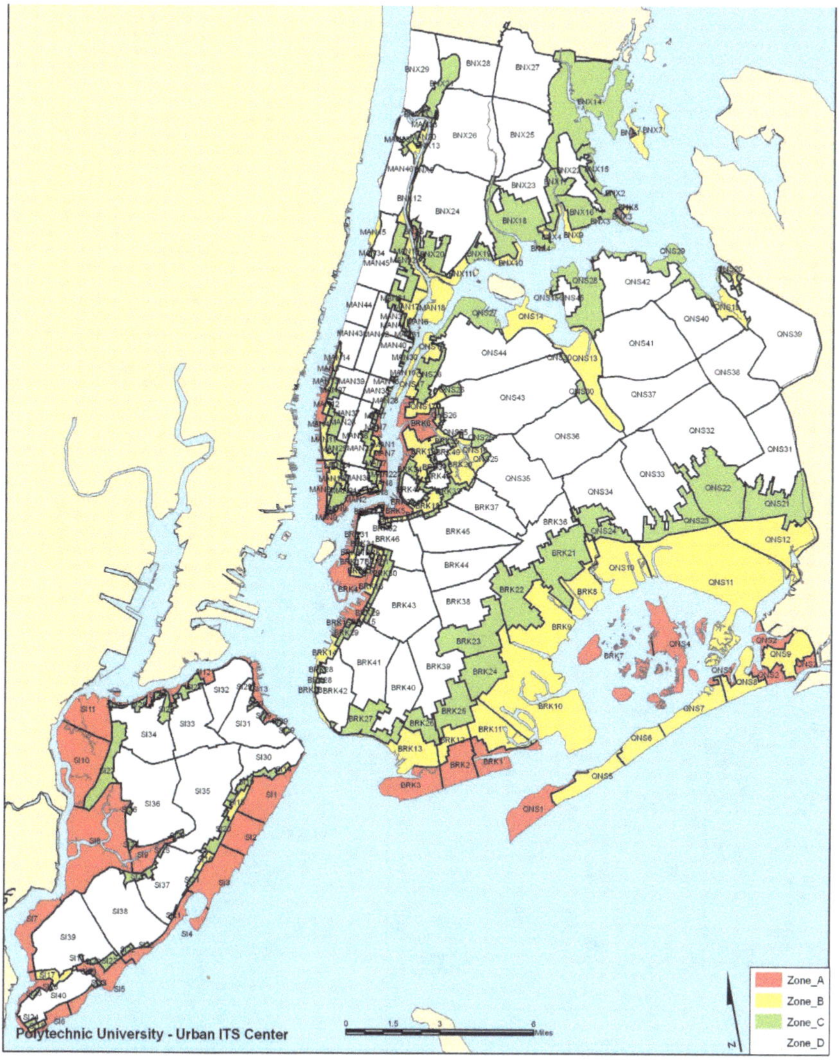

FIGURE 12-2 Evacuation zones in New York City.

criteria for the placement of highway advisories in light of emergency evacuation requirements. It is important, therefore, for transportation planners to consider the needs of emergency evacuation in locating traffic sensors and other information devices for traffic management. This can only happen when transportation planners work closely with emergency personnel in the planning and management of transportation systems.

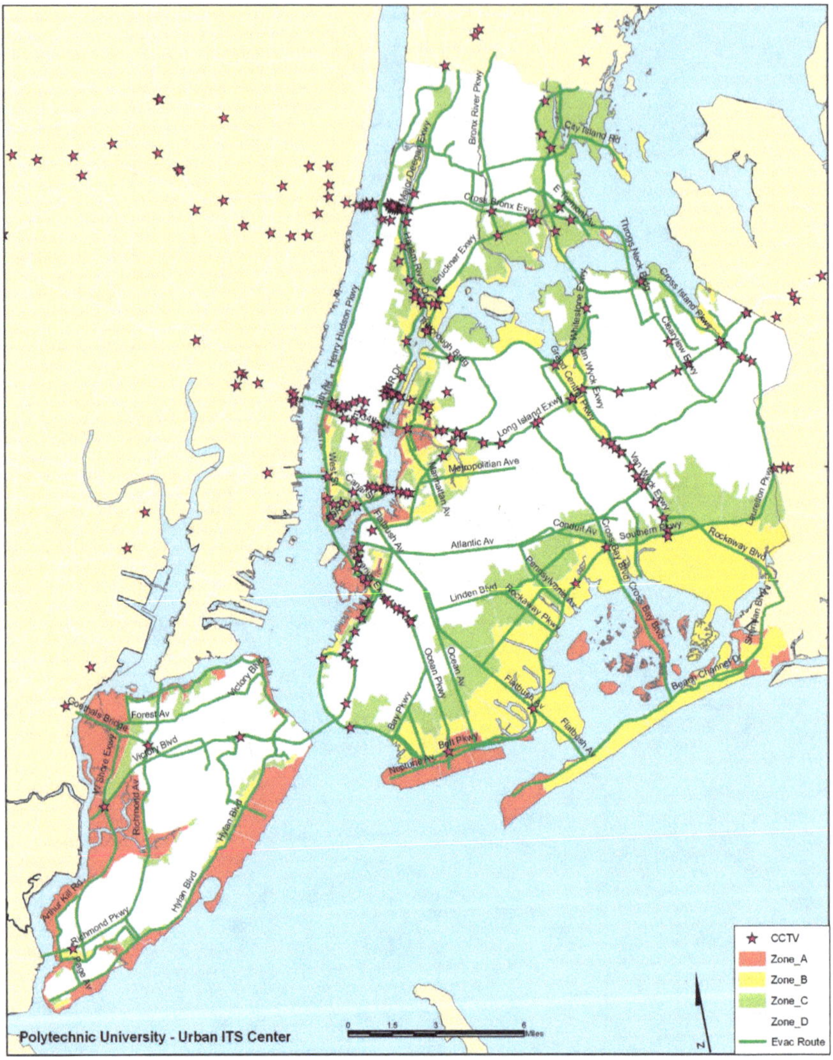

FIGURE 12-3 Existing CCTV locations.

Organizational Preparedness

When preparing emergency plans, many local governments have traditionally relied on their police departments' plans. Transportation agencies with extensive traffic management capabilities are not adequately involved with the police departments in the preparation of such plans. In most cities the police are not

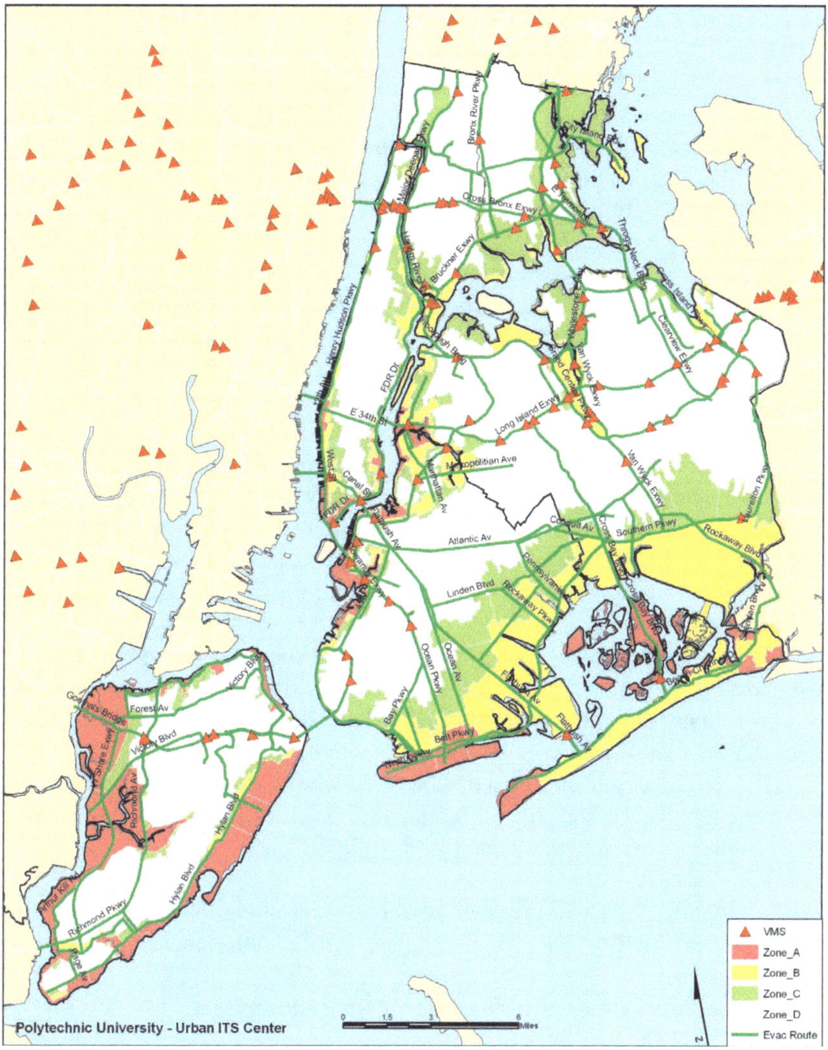

FIGURE 12-4 Existing VMS locations.

experts in transportation planning and traffic management and are generally not aware of the advanced technologies used by traffic engineers to control traffic. In fact, police officers rely on manual traffic control and tend to override the signal-timing patterns in place. Such lack of knowledge on the part of the police contributes to inefficiency in transportation response and unnecessary traffic delays to the evacuees.

In addition, the timely deployment of police officers to the affected areas requires an ability to provide transportation access to police personnel who live outside the city. The McKinsey report on the September 11, 2001, terrorist attacks in New York City suggests that a more effective mobilization of police officers to the rescue site is dependent on transportation.[10] For example, some 10,000 New York City police officers live on Long Island, and a thousand more first responders and other public employees also live far away.

The transportation agencies, on the other hand, are focused on system management for expected daily traffic patterns but are not closely connected with the emergency issues and procedures that the police use in managing emergencies. This lack of coordination results in a loss of efficiency that engineering applications would provide. An example of how transportation professionals can help in managing emergencies was demonstrated in New York City, where the Traffic Management Center was able to provide emergency-access-only lanes on the city's highways during the September 11, 2001, emergency.[11]

In addition to the need for integrating police and transportation personnel in emergency response functions, it is also necessary to coordinate the use of transportation resources with those of adjoining political jurisdictions as traffic from the movement of evacuees spills outside the disaster area (for example, Hurricane Katrina). Table-top and field exercises are two of the most effective strategies to achieve this objective.[12]

Therefore, since emergencies are infrequent events, it is important for first responders to be ready to function in a coordinated way when they are needed. However, many jurisdictions and agencies have developed emergency response plans that lack in their details for implementation.

Emergency action plans need to specify who will do *what* and *when*, and they need to establish the chain of command for different types of emergencies. An example of some recommended steps follows:

- Identify who will provide special services along the evacuation routes. These services include water, fuel supplies, information, medical services, and vehicle repairs.
- Coordinate the above functions in time and space.
- Establish clear lines of command.
- Create communication networks that provide support to those who need assistance because they are infirm, cannot drive, or do not have access to private transportation.
- Establish a current inventory of people who may need assistance and how to contact them in an emergency.

The above guidelines will go a long way in addressing the problems reported in Houston by a recent *New York Times* article, in which it was noted that technology is easier to install than to use effectively, as this depends on the ability of coordinating the activities of the responding agencies as well as ensuring the reliability of the equipment over time. In Houston, when tularemia was detected in October 2003, "equipment was installed quickly, but there was no detailed plan in place for how to respond to positive alarms."[13]

EVACUATION PHASE

The effectiveness of an evacuation plan can be measured by the time it takes to safely evacuate the population from the disaster zone. The critical factor affecting this outcome is maintaining the *capacity* of the evacuation routes.

Traffic Monitoring During Evacuation

Monitoring of traffic conditions (time, location, and duration of traffic incidents; location of traffic bottlenecks and queues) during evacuation will allow for prompt response to correct anomalies emerging from the experience. Perhaps even more important is the task of protecting and sustaining the capacity of the network at critical locations. Some of the more critical steps that should be taken include the following:

- **Posting tow trucks along evacuation routes for incident-free evacuation**: Appropriate measures must be taken to allow incident-free traffic flow along evacuation routes. To preserve and restore roadway capacity, all incidents must be quickly handled and lanes must be cleared for unimpeded traffic flow.
- **Access control**: Highway ramps play a critical role in the transportation network, and their efficient operation is vital to system capacity. Therefore, utilizing redundant control devices and enforcement at on- and off-ramps along expressways and major highways are key requirements to maximize the use of available capacity.
- **Ramp management**: Closing ramps at critical locations along expressways and highways may be necessary during an evacuation to ensure maximum capacity on the evacuation routes. The transportation agency and the police should coordinate ramp location policies along the evacuation routes.
- **Bottleneck situations**: To avoid gridlock, critical intersections, ramps, and approaches should be posted with traffic control agents to facilitate traffic and lane management.
- **Suspension of work-zone activity**: Roadway capacity is also affected by ongoing work-zone activities. At minimum, work zones should be suspended and equipment should be removed from roadsides and shoulders during the evacuation and recovery periods.
- **Media advisory and public outreach for evacuation routes**: A media advisory for public outreach should contain a message that the evacuation routes are priority routes and nonevacuees should stay away from such routes while

ordered evacuation is in progress. This will allow authorities to control access to evacuation routes.

Monitoring and documenting the results of a response operation will provide the basis for updating the plan in light of lessons learned. This activity will ensure that future responses will be more effective than earlier ones.

CONCLUSIONS

Effective evacuation planning is dependent on a functioning transportation system during emergencies. Such a requirement may only be met if we rethink some of the traditional practices for transportation system planning and create a collaborative environment among transportation planners and first responders and between various levels of governments and jurisdictions. Finally, detailed plans for evacuation need to be developed, tested, and updated to reflect lessons learned through firsthand experience or through experience transferred from other cities.

NOTES

1. Federal Highway Administration, Office of Legislation and Intergovernmental Affairs, Program Analysis Team. 2005. A Summary of Highway Provisions in SAFETEA-LU. Available online at *www.fhwa.dot.gov/safetealu/summary.htm*. Accessed April 24, 2008.
2. Howitt, A. M., and J. Makler. 2005. On the Ground: Protecting America's Roads and Transit Against Terrorism. The Brookings Institution Series on Transportation Reform. Washington, D.C.: The Brookings Institution. Available online at *www.brookings.edu/~/media/Files/rc/reports/2005/04transportation_howitt/20050426_howitt.pdf*. Accessed April 24, 2008.
3. American Society of Civil Engineers. 2006. ASCE News 31(9).
4. See the Web site of the Intelligent Transportation Systems Program of the U.S Department of Transportation at *www.its.dot.gov/index.htm*. Accessed April 24, 2008.
5. Center for Transportation Studies, University of Minnesota. 2002. How Should Transportation Change After September 11? Summary report of the Inaugural James L. Oberstar Forum on Transportation Policy and Technology. Available online at *www.cts.umn.edu/Events/OberstarForum/2002/documents/2002oberstarforum.pdf*. Accessed April 24, 2008.
6. Litman, T. 2005. Lessons from Katrina and Rita: What major disasters can teach transportation planners. Journal of Transportation Engineering 132:11-18. Also presented at the 85th Transportation Research Board Annual Meeting, January 22-26, 2006, Washington, D.C. Available online at *www.vtpi.org/katrina.pdf*. Accessed April 24, 2008.
7. Howitt and Makler. Protecting America's Roads and Transit.
8. Patel, R. K., and J. C. Falcocchio. 2005. An improved managed lane framework for emergency management. Paper presented at the 85th Transportation Research Board Annual Meeting, January 22-26, 2006, Washington, D.C.
9. Urban Intelligent Transportation Systems Center, Polytechnic University. 2006. Intelligent transportation systems plan for hurricane evacuation, Task 1, technical memorandum prepared for the New York City Office of Emergency Management, October 12, 2006.
10. McKinsey & Company. 2002. Improving NYPD emergency preparedness and response. Available online at *www.mipt.org/pdf/nypdlessonslearned9-11.pdf*. Accessed April 24, 2008.

11. Tipaldo, J. M. 2003. 9/11 and New York City's traffic management center: Before, during, and after. Presentation at the New York Chapter Meeting of the Institute of Transportation Engineers, Sarasota, N.Y., July 2003; Talas, M. 2003. Maintaining highway mobility during emergency: NYC highways implementation post September 11. Presentation at the Annual Meeting of the Institute of Transportation Engineers, Seattle, WA, August 2003.

12. Ritter, L., M. J. Barrett, and R. Wilson. 2007. Securing Global Transportation Networks: A Total Security Management Approach. New York: McGraw-Hill.

13. Lipton, E. February 9, 2007. New York to test ways to prevent nuclear terror. Online. The New York Times. Available at *www.nytimes.com/2007/02/09/nyregion/09nuke.html*. Accessed April 25, 2008.

13

International and National Priorities in Combating Terrorism in the Transportation Sector

Vladimir N. Lopatin,
Republic Scientific Research Institute of Intellectual Property

Despite the increasingly systemic and organized efforts of the international community and individual states, terrorist threats continue unabated. In an effort to inflict the most serious damage and intimidate the government and people, terrorists select the most vulnerable targets for their attacks.

BACKGROUND

To strengthen the fight against terrorism and improve its effectiveness, 7 years ago we Russian scientists concluded that it made sense to focus antiterrorist activities not only on critical areas particularly dangerous from the standpoint of the threat of terrorist attacks but also on especially critical facilities. These primarily include transport, which, because of its transnational nature, serves as (1) an environment for "economic" activities of international terrorist and other criminal groups, (2) a target for banditry, or (3) a means for the perpetration of terrorist acts. Although up to 70 percent of terrorist acts are either committed on transport or involve its use, this has not been reflected adequately in legislative or law enforcement practice.

Following analysis of the situation in Russia and in the Commonwealth of Independent States (CIS) as a whole both at the international conference "Terrorism and Transport Security" held in Moscow on February 5-6, 2002, and subsequently, experts identified a number of reasons why transportation may be categorized as a critical target:

- Sharp increase in hazardous cargo as a proportion of the total volume of goods transported
- High level of infrastructure decay and high accident rate in the transport sector
- Relative accessibility
- Use of smuggling by transnational criminal groups as a source of financing for terrorism
- Possibility of attracting broad public and media attention
- Association with national symbols (national airlines)
- Possibility that even a single act or attack will immediately affect many people

For these reasons, security and crime prevention in the transport sector is one of the priorities of the state and society.

Based on an analysis of legislation and law enforcement practices in 2000-2001, it was clear that transport policy did not include an antiterrorism component, and antiterrorism activities did not focus on transport. There was very little overlap between these two sectors. At that time we were asked to define a priority area in transport policy, namely, ensuring security and antiterrorism activities, and to create a similar focus on the transport sector as a priority area in antiterrorist activities. The CIS Transportation Coordinating Council agreed with this assessment in the Chisinau Declaration on Transportation Safety, as did the Council of Ministers of the European Conference of Transportation Ministers in its closing document and in the Bucharest Declaration on Combating Terrorism in Transport on June 6, 2002. This approach is also reflected in a statement on the fight against terrorism in the transport sector adopted on June 28, 2002, at the summit of the Group of Eight (G-8) in Kananaskis, as well as in subsequent decisions of international and state structures. Further evidence that the first steps have been taken in developing this consciousness and understanding may be seen in subsequent years when the first intergovernmental agreement on transportation safety was adopted, including a set of principles and mechanisms for implementing state policy on transportation security and counterterrorism. Another important step was the adoption in 2006 of the Federal Law on Transportation Security. A comprehensive and systematic approach to these problems is reflected in the CIS Intergovernmental Program of Joint Measures for Combating Crime in 2005-2007 and in cooperative programs among CIS member states for fighting terrorism and other extremist phenomena in 2005-2007 and countering the illicit drug trade.

Thus, the problem of understanding and awareness at the level of experts, academics, and individual government officials and business leaders has today reached the level of government and international understanding, which has been reflected in specific decisions made by state authorities and international organizations in this regard. These decisions may be considered a starting point in improving transport and antiterrorism policies and in creating on this basis an integrated new sphere of state policy: transportation security and counterterrorism.

The new approach in the formation of the counterterrorism strategy and its implementation in the transport sector is complex both in its identification of the targets and the subjects of counterterrorist activities and in the principles and mechanisms of their interaction. And today, scientists must help to take the next step in implementing this approach to promote security and counterterrorist objectives in the transport sector.

FOCUS AREAS FOR ANTITERROR ACTIVITIES

Comparing the norms of international laws to which Russia is a party with norms of Russian legislation, it should be recognized that there is currently no antiterrorist strategy on transport that would be mandatory for state structures, including both the transport complex and law enforcement, and that would take into account the specifics of all types of transport, from aviation to subway systems. It would probably be wrong even to implement a counterterrorism policy at all without making it specifically applicable to elements of the transportation sector, given its very substantial special characteristics.

An analysis of international agreements on counterterrorism suggests that of the seven major transportation modes, air and sea transport are subject to the most restrictions. The other modes are either only the subject of general mention and declarative statements or entirely absent from the list of antiterrorist activities. For example, the Concept for a Coordinated Transport Policy among CIS member states for the period through 2010, approved by decision of the Council of CIS Heads of Government on September 15, 2005, mentions only aviation, marine, river, and rail transport with regard to antiterrorism and security activities related to transport policy. Such a mention is lacking in the section on vehicular transport, while pipelines and subways are not included in the document at all.

Therefore, it remains an urgent challenge to adopt an antiterrorism strategy for the transport sector to ensure transportation security as an integral part of the international counterterrorism system. This will entail developing and making the necessary amendments to the Transport Strategy and to targeted programs for modernizing the transport system. It will also require securing allocations in the antiterrorist strategy for law enforcement to make counterterrorism on transport a priority, taking into account the specific characteristics of its seven major modes.

ORGANIZATIONS INVOLVED IN ANTITERROR EFFORTS

In addition to law enforcement, other government agencies and nongovernmental organizations must be key actors in carrying out crime prevention measures to implement United Nations Security Council Resolution 1373 of September 28, 2001, as without them this effort will be ineffective and will not produce the expected results. Preventing terrorism can and must be done through the joint efforts of all government agencies and with the support of civil society, science, and business at both the national and international levels, including in the CIS. The new conditions require new rules for interaction between government, science, and the business community in order to establish partnerships in addressing the common task of countering terrorism.

In 2001, Russian scientists found that to confront the well-armed, well-trained, and highly professional enemy that is international terrorism, it is necessary not only to combine the efforts of the various law enforcement agencies but also to promote cooperation between law enforcement and transportation agencies; the state and nonstate sectors; and government, science, and business. This initiative to unite the efforts of government, science, and business led to the establishment on May 20, 2002, of the High-Level Advisory Group on Countering Terrorism in the Transport Sector in the Russian Federation. This group includes designated representatives of State Duma committees, all transport and law enforcement agencies, transport companies, the Russian Academy of Sciences, the Russian Chamber of Commerce and Industry, the CIS Transportation Coordinating Council, and the International Road Transport Union (IRU). The following accomplishments have been made on the initiative of the group and thanks to the efforts of researchers and practitioners:

- Unique counterterrorism experience, both negative and positive, has been summarized at six international conferences on terrorism and transportation security, the results of which have been published in individual book form. In 2006, by decision of the CIS Interparliamentary Assembly, the conference was given the status of a permanent CIS advisory body on counterterrorism and transportation security.
- Recommendations of the first five international scientific conferences on terrorism and transportation security are for the most part being implemented and are finding support in the decisions of CIS intergovernmental agencies, including the Interparliamentary Assembly, the CIS Executive Committee, state agencies of CIS member countries, and the Collective Security Treaty Organization. In particular, the recommendations of the third and fourth international conferences have formed the foundation for practical efforts to protect civil aviation against acts of unlawful interference.
- A unique set of statistics was collected on terrorism in the transport sector in the Russian Federation, international and national legislation in this area

was analyzed, and in 2003 an unparalleled white paper entitled "Terrorism and Transportation Security in Russia (1991-2002)" was published.

- A list was prepared of suspicious International Road Transport (TIR) carnet (shipment log) transactions that, if encountered, should cause national road shipment associations not only to refuse to issue (withdraw) a TIR carnet but also to inform relevant law enforcement agencies (similar to efforts established to counter money laundering). In December 2002 the list was adopted as a regulation at the IRU General Assembly in Geneva, requiring compliance by all national road shipment associations in the 64 IRU member countries.

This Russian initiative was supported in the Chisinau Declaration on Transportation Safety (May 27, 2002), which was adopted at a meeting of the CIS Transportation Coordinating Council featuring the participation of all ministers of transport from the CIS member. This declaration was circulated as an official document at the European Conference of Ministers of Transport, the World Road Transport Forum (June 2, 2002), and a joint session of the CIS Interparliamentary Assembly commissions on political affairs and defense and security in the city of Astana (October 24, 2002). It was also submitted as part of the G-8 Action Plan on a Secure and Facilitated International Travel Initiative (June 11, 2004). In particular, the September 18, 2003, decision of the CIS Council of Heads of Government directly orders law enforcement agencies to work with high-level advisory groups on transportation security.

A number of issues remain unresolved, including clear definition of objectives, functions, and structures; coordination and interaction of governmental and nongovernmental entities involved in countering terrorism in the transport sector; harmonization of regulations and the activities of state security structures; and implementation of measures to ensure the efficiency and optimal utilization of mobile transport units in counterterrorism activities. In particular, it is necessary to develop and establish in regulatory form mandatory procedures for cooperation among law enforcement agencies, national associations of road transporters, and transport organizations if they detect a suspicious operation, a list of which has been drawn up and approved by both the Russian Federation and the IRU. This experience and the favorable response that this Russian initiative has elicited suggest that further progress would be possible in setting priorities for international cooperation in the fight against terrorism in the transport sector.

PRINCIPLES FOR ANTITERRORIST ACTIVITIES IN THE TRANSPORT SECTOR

Along with the well-known and generally accepted principles of cooperation, the new conditions have given rise to new rules that remain to be adopted and established in regulatory form as uniform and compulsory for all participants in antiterrorist activities at both the national and the international levels.

The first issue is how to determine the balance between the interests of development (including freedom of movement) and the interests of security, the balance between obligations to provide protection against terrorist acts and the obligation to protect human rights.

With regard to striking a balance between ensuring human rights and combating terrorism, the Guiding Principles of the Council of Europe affirm a restriction forbidding arbitrary treatment and legislation with a retroactive effect, assert the right to a fair judicial hearing, and reject extradition of individuals to countries where they may be condemned to death. This is necessary but clearly not enough.

The strategy must be aggressive and specifically set forth in legal decisions. An aggressive strategy involves making adjustments and very substantial ones in the legislative framework, including standards and principles of international law that have long run counter to today's situation and have become obsolete, beginning with the Tokyo Declaration of the 1960s and other documents adopted regarding the transport sector in the 1970s. The situation has long since changed. Absolutely new threats have arisen, and it seems to me that understanding and awareness of this will only allow us to work together to adjust international legal standards and principles in order to counter our common enemy, international terrorism, and to amend national legislation as well, thus providing a sound foundation for the offensive against terror in the transport sector both in Russia and in the world as a whole.

Second, applying experience in counterterrorism activities to other modes of transport should be done in a gradual and rational fashion. For example, efforts are under way to resolve the issue of creating a technical monitoring system that will automatically collect and format data on suspicious signs suggesting preparations for commission of a terrorist act on transport. This system would include real-time transfer of data on passengers (passport data) from all transport enterprises regardless of their form of ownership to internal affairs agencies when passengers check in for air, rail, or water travel.

Third, in the fight against terrorism it is important to be consistent in keeping and using what works. An example is the rejection of participation by internal affairs agency personnel in joint predeparture inspections along with airline security personnel and the subsequent recognition that this decision had been a mistake. A similar error was made in eliminating the presence of personnel from the public prosecutor's office on transport and at sensitive facilities, a mistake that is being rectified by the new attorney general of Russia.

Fourth is the issue of developing and implementing common standards for security, for example, standards for installing modern equipment at inspection points at airports and seaports, including devices capable of detecting explosives on the human body, thus eliminating the need for manual searches. This rule is especially true, given the many structures created and operating in this sphere (more than 30 international entities operating in Russia).

International Counterterrorism Structures

- Counterterrorism Committee of the UN Security Council (established under UN Security Council Resolution 1373 of September 28, 2001)
- Interpol
- World Customs Organization
- International Atomic Energy Agency
- International Civil Aviation Organization
- International Maritime Organization

In Europe

- Action Against Terrorism Unit of the Secretariat of the Organization for Security and Cooperation in Europe
- Europol
- European Conference of Ministers of Transport
- Eurasian Transport Union

In Asia

- UN Economic and Social Commission for Asia and the Pacific
- Eurasian Transport Union

In the CIS

- Counterterrorism Center of CIS Member States
- Coordinating Office for Combating Organized Crime and Other Dangerous Types of Crime in CIS Member States
- Coordinating Conference of Attorneys General of CIS Member States
- Council of Heads of Security Agencies and Special Services of CIS Member States
- Council of Ministers of Internal Affairs of CIS Member States
- Council of Ministers of Defense of CIS Member States
- Council of Commanders of Border Forces of CIS Member States
- Council of Heads of Customs Services of CIS Member States
- Council of Ministers of Foreign Affairs of CIS Member States
- Transportation Coordinating Council of CIS Member States

Fifth is the issue of the transition from departmental and then program-based planning to targeted project-oriented planning, financing, and management, including for antiterrorist activities in the transport sector. This requires the development of a system of indicators regarding the main subjects of antiterrorist activity. Upon analysis of the summary report on the results and main activities of the government of the Russian Federation for 2006-2008, it is possible to state that such a system does not exist (the report's section on transportation security features only two indicators for assessing the activities of government agencies regarding air transport). The task of scientists is to help the authorities detect and recognize problems and find ways of solving them. To ensure that scientific recommendations form the basis for improvements in the future work of state and nongovernmental structures and organization of interactions between them, it is important to follow the principle of joint operation, working together instead of trying to replace one another.

14

Managing the Radius of Risk

Lieutenant Colonel Drew F. Lieb,
Deputy Superintendent of Homeland Security, New Jersey State Police

Energy infrastructure security in the state of New Jersey is taken as a very serious matter. The New Jersey State Police (NJSP) took innovative steps after September 11, 2001, to address this issue and to blaze new paths to detect and prevent any further terrorist attacks. The Homeland Security Branch of the State Police is achieving this goal with their philosophy of "All crimes, all hazards, all threats, all the time."

Following the September 11, 2001, terrorist attacks, the Nuclear Regulatory Commission (NRC) immediately advised nuclear facilities to go to the highest level of security in accordance with the system in place at the time. Advisories, orders, and guidance documents have since been issued to further strengthen security at nuclear power plants. While specific actions taken remain sensitive, they generally include increased security patrols, augmented security forces, additional security posts, installation of additional physical barriers, vehicle checks at greater stand-off distances, enhanced coordination with the law enforcement and intelligence communities, and more restrictive site controls for all personnel. As a result, some state governors assigned National Guard troops and state law enforcement agencies to work with security forces at nuclear power plants in their states. And the NRC continually assesses the threat environment in coordination with federal, state, and local law enforcement agencies. The NJSP has an ongoing security initiative with all the nuclear electrical plants within the state.

These initiatives add to the security posture already in place at these plants. This initiative is not just a force multiplier; it also provides for a system of additional monitoring at a statewide level. The colossal scale of pipeline and electrical infrastructure in the United States alone—more than 160,000 miles of crude oil pipelines, 4,000 offshore platforms, 10,400 power plants, and 160,000 miles of transmission lines—makes providing security a daunting challenge.

Terrorist attacks, in particular, pose a grave threat. In videotape released in December 2005, deputy al Qaeda leader Ayman al-Zawahiri singled out energy infrastructure as a key strategic target for his followers.

Drive up and down the New Jersey Turnpike, and it is easy to see why this state is a potential playground for terrorists. There is a 2-mile stretch from Newark Airport to Port Elizabeth that terrorism experts have called "the most dangerous 2 miles in America."

A corridor state between Washington, D.C., and New York City, New Jersey is no stranger to terrorists or their acts. Whether it was the violent domestic terrorist groups of the 1960s and 1970s or the international terrorist cells of the 1980s and beyond, New Jersey has played host to their presence and their attacks. Because of the vast diversity of the state's population, multinational terrorists have easily blended into the urban populations and assimilate within the ethnic cultures, while planning and eventually executing attacks against the major target across the Hudson River, New York City.

New Jersey is the most densely populated state in the country. On this particular swath of land there are hundreds of potential terrorist targets—chemical plants, rail yards, rail lines, refineries, pipelines, an international airport, and the third-largest port in the United States. In a worst-case scenario, the potential to bring harm to more than 12 million people lies within a 14-mile radius. New Jersey comprises 21 counties with a population of 8.5 million; this is the highest population density of any state in the United States. With an average of 1,135 people per square mile, New Jersey's population density is 13 times the national average. The Port of Newark–Elizabeth Marine Terminals are one of the world's largest container ports. Newark International Airport is ranked seventh among the nation's busiest airports and is the fifth-busiest international air gateway into the United States. Additionally, New Jersey is home to the largest petroleum containment system outside the Middle East. With a dense population and a major industrial base, the protection of New Jersey's citizens and critical infrastructure is a top priority of the NJSP.

The terrorist attacks on September 11, 2001, confirmed that all Americans share responsibility for homeland security. Federal, state, local, private-sector, and nongovernmental entities and individual citizens across the state of New Jersey and the nation need to prepare as one entity for major events that may exceed the capabilities of any single agency. The American structure of overlapping federal, state, and local levels of governance provides unique opportunities and challenges. Opportunities arise from the flexibility to explore differences, based on unique roles and responsibilities, and share best practices from across the state and nation. Challenges arise from the need to develop interconnected and complementary state and national homeland security strategies that respect those differences and balance flexibility with accountability.

HISTORY

The catastrophic events of September 11, 2001, forced the NJSP to undergo a paradigm shift in its overall approach to its duties and responsibilities. This horrific act was the catalyst for a major cultural change in the capability of law enforcement to provide a "proactive preventive defense" to thwart a terrorist event. The ability of the NJSP to sustain a visible police presence during elevated alerts placed an enormous strain on all assets and resources. On February 26, 2004, the NJSP announced the largest reorganization in its history with the creation of the Homeland Security Branch (HSB). The branch was conceived as two separate entities within one command structure—on one side, the Special Operations Section to provide both an immediate and a sustainable response and, on the other, the Emergency Management Section (EMS) immersed in the preparation and mitigation of any realized event within New Jersey.

This reorganization fostered the adoption of a regional concept in an all-hazards approach to thwarting terror events, providing an immediate response mechanism, mitigating any natural disasters, and reducing crime. The pressures on our resources and assets have not been greater and will continue to grow in the years ahead. Any law enforcement response must be a concerted effort with a central focus on an efficient reality of addressing today's fiscal climate. HSB must remain cognizant in properly managing its resources and assets while providing a sound proactive preventive defense for New Jersey. The ability to provide an effective response mechanism in an all-hazards approach to any realized natural or terrorist event will position New Jersey to prevent acts of terrorism, protect critical infrastructures and key resources, and mitigate any type of disaster.

The NJSP has developed a multifaceted approach to interconnected and complementary homeland security strategy. First, the NJSP has formulated strategies in accordance with and under the direction of the U.S. Department of Homeland Security's National Response Plan and National Priorities. Under the direction of the latter, NJSP HSB makes use of NJSP's intelligence-led policing strategy to ensure that the necessary essential risk management assessments are conducted, all threats and vulnerabilities are identified, and the appropriate response is initiated.

Once a potential threat or hazard has been identified, HSB is charged with gathering and documenting incident-related facts for recovery efforts and lessons-learned analysis. HSB's mission is to provide a continuing level of preventive security and public safety through the efficient utilization of statewide resources.

HSB also employs community policing strategies to gather information and intelligence. Proactively and continually, HSB personnel maintain active two-way communications with the public at-large, where information and intelligence is obtained. Importantly, specific and relevant information is returned to the community. It is the intent of HSB that the community is aware that their contributions to homeland security are evaluated and their efforts are appreciated. It is also the intent of HSB to provide the community with appropriate training and equipment to aid in the homeland security mission. Infrastructure and force protection concerns can be immediately addressed through the Regional Operations Intelligence Center (ROIC) and an appropriate response can be directed. The Maritime Security Initiative (MSI) is one venue that HSB uses to interact with the community. Funding for equipment and training for preventive security is administered through the Urban Area Security Initiative (UASI). UASI has been designated to provide immediate access to specialized assets within the state's most populated region.

Second, NJSP has reorganized its structure to place all existing front-line defenses under one unified command to more effectively and efficiently respond to calls for police service that involve the safety of citizens and critical infrastructure within the state and any large-scale national disasters. As a result of NJSP's reorganization the HSB was created. A more detailed restructuring of NJSP placed the Special Operations Section (SOS) and the EMS under the direct command of HSB. The SOS was formed to act as NJSP's rapid-response, all-hazards force. Under SOS, all existing frontline personnel and equipment were placed under one unified command. The SOS objective is to provide an increased and diverse presence while responding to a critical incident within the state or an out-of-state incident of national significance. If either an incident of national significance or a critical incident affecting the state occurs, SOS is prepared to mobilize personnel and resources in concert with our law enforcement, emergency response, and private-entity partners. NJSP and specifically SOS personnel are now better prepared to respond to any situation that the state or nation may face. Operation Louisiana Emergency Assistance Deployment (LEAD) demonstrated to the nation how NJSP's and the state's public safety resources could be effectively deployed by responding to the Gulf Coast areas affected by Hurricane Katrina.

All sworn NJSP personnel and many civilian personnel assigned to HSB are highly trained in very specialized law enforcement-related fields. Specialization includes, but is not limited to, special weapons and tactics, explosives, hazardous materials handling, commercial vehicle safety, traffic accident reconstruction, aviation, maritime-related initiatives, and governmental infrastructure security.

MISSION STATEMENT

The mission of HSB is to provide a proactive, preventive defense regarding critical infrastructures and key resources through a sound intelligence-based collaboration with all our public- and private-sector partners, utilizing a philosophy of the regional design concept. The terrorists and criminals do not respect geographic, political, or legal jurisdictional lines and often exploit these boundaries. Similarly, the threat from natural disasters will continue to threaten the lives and safety of all New Jersey citizens. This proactive philosophy will aid in thwarting any asymmetric threats and provide for the proper response and mitigation to any realized natural disasters within the state of New Jersey.

The Special Operations Section consists of six separate and distinct bureaus: (1) State Governmental Security, (2) Marine Services, (3) Aviation, (4) Transportation Safety, (5) Deployment Services, and (6) Technical Response. The Emergency Management Section consists of three separate and distinct bureaus: (1) Communications, (2) Emergency Preparedness, and (3) Recovery.

SPECIAL OPERATIONS SECTION

State Governmental Security Bureau

The State Governmental Security Bureau organizes, directs, staffs, coordinates, and reports the activities of the Security Operations Unit, State House Complex Security Unit, Justice Complex Security Unit, Investigations Unit, Central Security Unit, and the Executive Protection Unit. This entity facilitates the flow of information to and from the various units supervised and serves as a conduit for communication with other division entities. The chief of the State Governmental Security Bureau serves as the superintendent's representative on the Capitol District Oversight Committee and the State Government Operations Group Committee under the umbrella of the Domestic Security Preparedness Task Force. The bureau also processes and issues permits to gather or use amplification equipment in or around state-regulated buildings and grounds. The State Governmental Security Bureau is committed to providing security and protection to visitors, employees, and property within the State Capital Complex. These critical services are provided in a professional, unbiased, and courteous manner. Realizing the importance of building agency partnerships, bureau members are dedicated to the concept of service-oriented policing. The bureau proudly preserves the traditions of the state police by "maintaining the good opinion of the people of the State of New Jersey."

Marine Services Bureau

The New Jersey State Police Marine Services Bureau (MSB) is the primary provider of full-time law enforcement services for more than 200,000 registered vessels on all of New Jersey waterways and contiguous land areas. The mission of the MSB is to protect and serve our citizens and every aspect of the marine environment, preserve natural resources, enforce the laws of this state, and provide a preventive measure of homeland security that is second to none.

The Marine Services Bureau comprises five main stations and four substations. These facilities are strategically located throughout the state to address recreational boating issues, fish and game laws, search and rescue, criminal matters, and homeland security. The stations are located in the following areas:

- Atlantic City
- Bivalve
- Burlington
- Lake Hopatcong
- Monmouth County
- Newark Bay
- North Wildwood
- Ocean (Waretown)
- Point Pleasant

The waters of this state include the following:

- 1,960 square miles of coastline, fresh water lakes, and rivers
- 127 miles of Atlantic Ocean coastline
- 1,750 miles of interior tidal shoreline
- 100 inland bays, creeks, coves, and rivers
- More than 800 lakes and ponds totaling more than 700 square miles of surface area

Supplemental to standard state police training, all of the troopers assigned to the Marine Services Bureau patrol function attend a 4-week marine law enforcement school and then must demonstrate their proficiency through successful completion of the Vessel Operator Certification Program. To maintain a high level of proficiency, the certification process must be revalidated annually.

Separate from and supplemental to internal certifications, approximately one-third of the personnel assigned to the Marine Services Bureau are captains licensed by the U.S. Coast Guard, with licenses that include both tonnage and commercial towing endorsements. Marine Services Bureau training also includes water survival, ocean rescue, and ice rescue.

Patrol vessels vary widely from 13 to 50 feet in length, from single outboard to twin diesel inboards producing in excess of 1,000 horsepower, from fiberglass to aluminum, and from open to fully enclosed weather-tight cabins with long-range capabilities. The equipment onboard the vessels include basic marine safety equipment, very high frequency (VHF) radios, police radios, high-technology thermal imaging equipment, side-scan sonar, depth finders, and radar-interfaced navigation equipment.

As a preventive measure intended to increase boating safety and reduce waterway user conflict, during the winter months a contingent of troopers is assigned to various schools throughout the state to teach boating safety to students.

Aviation Bureau

The Aviation Bureau provides emergency medical evacuations (medevac) transportation of seriously injured victims of motor vehicle, industrial, recreational accidents, and so forth, to trauma centers. It also provides air support for the various commands within the Division of State Police, as well as other law enforcement agencies that request assistance, in accomplishing numerous police and homeland security activities. It is the responsibility of the Aviation Bureau to provide services for the following:

- On-scene medevac transportation of seriously injured victims of motor vehicle, industrial, and recreational accidents, and so forth, to trauma centers
- Interhospital medevac transportation of seriously ill patients to specialty care facilities, such as burn centers, reimplantation centers, cardiac centers, and so forth

The Aviation Bureau will provide air support for the various commands within the Division of State Police and other law enforcement agencies that request assistance in accomplishing their police and homeland security mission. The bureau provides airborne expeditious search and rescue, including forward-looking infrared capabilities; aids disabled motorists; and facilitates traffic flow by identifying congested areas and suggesting solution alternatives. Aviation Bureau homeland security operations also include identification and surveillance of important infrastructure, including bridges, tunnels, power plants, refineries, and railways. Additionally, the Aviation Bureau will provide alert notification in selected areas of the state's Emergency Planning Zones and provide surveillance of evacuation areas in the Emergency Planning Zones.

The Aviation Bureau is responsible for maintaining its fleet of aircraft in compliance with all applicable federal aviation regulations, airworthiness directives, manufacturer's service bulletins, and aviation maintenance manual procedures and for ensuring that all Aviation Bureau maintenance technicians are properly trained and certified to maintain bureau aircraft in an airworthy condition at all times. The Aviation Bureau ensures that all pilots are properly trained and proficient by complying with Aviation Bureau performance standards as outlined in the operations manual and that all pilots meet Federal Aviation Administration (FAA) recency-of-experience requirements for night operation, instrument currency, and flight reviews.

Transportation Safety Bureau

The Transportation Safety Bureau (TSB) acts as the executive liaison to the Department of Transportation, Division of Motor Vehicles, Division of Highway Traffic Safety, and the Federal Highway Administration. The bureau provides technical assistance (including, but not limited to, commercial motor vehicle accident investigations) to state and municipal police departments, prosecutor's offices, the general public, and other government agencies. While the bureau's primary function is designated as commercial vehicle enforcement, TSB has a very important secondary role. The bureau acts as a rapid deployment force within HSB. The bureau is equipped to react to any state emergency and operate in a buffer-zone protection plan, providing a force multiplier in the detection of terrorist threats and actions in the all-hazards–all-crimes spectrum.

Commercial Carrier/Safety Inspection Unit

Commercial Carrier/Safety Inspection Unit personnel are responsible for implementing and enforcing federal regulations governing commercial vehicle drivers, related safety equipment, and the transportation of hazardous materials over state highways. They are also responsible for enforcing commercial vehicle size and weight laws. Having adopted the Federal Motor Carrier Hazardous Materials Regulations and the Federal Motor Carrier Safety Regulations, the division has assigned numerous teams of specially trained troopers to conduct roadside inspections of commercial vehicles to enforce federal safety regulations. Additional responsibilities include unannounced school bus safety inspections and commercial vehicle safety presentations to both the public and the private sectors.

Construction Unit

The Construction Unit enforces the rules and regulations governing traffic control and safety in highway work areas. The unit's members inspect New Jersey Department of Transportation (NJDOT) construction sites to ensure that contractors are complying with the traffic control plans established for their projects. Troopers assigned to the unit receive specialized training in work-zone safety and traffic control for highway construction areas. This training is combined with their experience in motor vehicle law enforcement to create a comprehensive safety program. The unit's members also provide work-zone safety training for local police agencies and for other governmental and private organizations.

Diesel Emissions Unit

The Diesel Emissions Unit (DEU) works in conjunction with the Motor Vehicle Commission to conduct roadside emission testing of heavy-duty diesel trucks, buses, and other diesel-powered vehicles. DEU is responsible for implementation and enforcement of federal regulations governing commercial vehicle drivers and related safety equipment. The unit also is responsible for enforcing state statutes governing size and weight regulations.

Hazardous Material Transportation Enforcement Unit

Hazardous Material Transportation Enforcement Unit (HMTEU) personnel are responsible for roadside hazardous materials inspections as well as commercial vehicle inspections. HMTEU has the primary responsibility for enforcing Title 49 of the Code of Federal Regulations along with the Hazardous Material Regulations also defined in Title 49. HMTEU is also responsible for an overtime joint Federal/State Internal Revenue Service Dyed Diesel Fuel Program.

Motor Coach/Compliance Review Unit

Motor Coach/Compliance Review Unit (MCCRU) personnel are responsible for roadside inspections of motor coaches, buses, and other commercial vehicles. The unit enforces Title 49 of the Code of Federal Regulations and Title 39 of the state motor vehicle code. The MCCRU also maintains the New Entrant Safety Audit Program. This program is a Federal Motor Carrier Safety Administration (FMCSA) initiative that involves troopers meeting with representatives from motor carriers who have applied for a federal Department of Transportation (DOT) number. During the meetings, carrier representatives are informed of the minimum requirements needed to operate within the guidelines of the FMCSA.

The MCCRU conducts compliance reviews of motor carriers that have failed to maintain an acceptable safety rating or have been involved in a fatal or otherwise serious commercial motor vehicle crash. This review involves an extensive check of a motor carrier's records, equipment, and drivers. This is an enforcement program, which the FMCSA utilizes to impose fines and out-of-service orders. Additional responsibilities include instructing motor coach inspection courses. The courses are given around the country and are mandated by the unit's funding source. The unit is also responsible for responding to and assisting with postcrash inspections.

Deployment Services Bureau

The Deployment Services Bureau consists of the following units:

- Infrastructure Security Unit
- Events Planning Unit
- Incident Management Unit

The Infrastructure Security Unit provides professional and technical assistance to agencies through the development of security surveys, vulnerability assessments and buffer zone protection plans (BZPP). The unit serves as the coordinator for providing assistance to the Office of Homeland Security and Preparedness Critical Infrastructure BZPP site survey program. They evaluate and review existing security plans, providing recommendations for modifications and improvements. They assist in developing comprehensive security plans for demonstrations, protests, rallies, and major political events conducted in and around state government buildings and grounds, as well as other identified critical infrastructure.

The Events Planning Unit coordinates operational and administrative planning for events and incidents that require assets outside of troop operations deployments. Events Planning maintains a centralized file of all plans developed for events and incidents while conducting and preparing pre-action and after-action planning reports, providing recommendations to the deputy superintendent of homeland security/superintendent for consideration and information. The unit coordinates planning activities with the Emergency Management Section for planned events and actual incidents as well as activities and details with other federal, state, and municipal agencies. The unit serves as liaison and point of contact with the New Jersey National Guard. They coordinate and maintain State Police Emergency Event Deployment (SPEED) recall rosters with the assistance of division section administrative officers. They coordinate special details as a result of any homeland security initiative (that is, Target Hardening/THREAT, Level Orange Deployment).

The Incident Management Unit (IMU) serves as a member of the Incident Management Response Team (IMRT) and responds to intermodal transportation incidents and other incidents as dictated by the Special Operations and Emergency Management sections protocols. They serve as a liaison to the incident commander. They ensure that all management issues are satisfied, including, but not limited to, asset management, maintenance of operational time lines, and logistical and planning support with traffic routing. IMU is responsible for coordinating NJDOT engineering staff and federal, state, county, and local agencies in the development of detailed diversion plans for state and interstate highways. They work with local, county, state, and federal agencies and their leadership to promote statewide incident initiatives. This will include attendance at monthly traffic officers, emergency management, safety council, emergency medical, and fire services meetings to market, assist, and develop planning tools for effective incident management. They respond and support all New Jersey Task Force One (NJTF-1) Search and Rescue operations. The unit also provides incident management training through outreach efforts to authorities in both the public and the private sectors.

Incident Management facilitates and coordinates postincident response evaluations (PIRE), which are designed to evaluate the emergency response to incidents for improved incident response and practices pertaining to transportation. They assist in the development of comprehensive operational plans for major events that support and promote the safety and well-being of all participants and attendees while working with established Traffic Incident Management Planning Teams (TIMPT) in all counties to develop contingency plans and other related initiatives that support the goal of "Keep the Traffic Moving."

Technical Response Bureau

Today, in advancing its overall homeland security mission, the NJSP has begun to implement a transformation process to better allocate its finite resources toward addressing terrorism and natural and manmade disasters. In keeping with the strategies and established practices outlined in the *National Strategy for Homeland Security* and *The 9/11 Commission Report*, HSB has evolved into a mission-oriented entity capable of confronting the challenges associated with emergency preparedness in a homeland security era. The creation of the Technical Response Bureau (TRB) placed those technical entities under the command structure of a single authority, enhancing our ultimate responsibility in maintaining the safety of our state and strengthening our homeland security mission. The TRB is an intricate component of HSB and is the primary technical response element for statewide emergencies. The bureau comprises four distinctive units:

1. Hazardous Materials and Response Unit (HMRU)
2. Arson/Bomb Unit
3. Technical Emergency and Mission Specialists (TEAMS) Unit
4. Canine Unit

The TRB instituted a capabilities-based strategy among its four subsidiary units. The capabilities-based strategy provides a framework for properly planned, organized, equipped, and trained personnel. Each unit maintains proficient capabilities within its respective discipline. However, cross-training among the units allows for a bureau-wide response to any critical event. This vital component of the TRB makes it the premier technical response entity in the state.

Hazardous Materials Response Unit (HMRU)

HMRU has dual areas of responsibilities within its mission. The unit provides operational response and planning support for force protection and for chemical, biological, radiological, nuclear, and explosive (CBRNE) incidents to include CBRNE agent surveillance and detection, identification of CBRNE material, evidence collection, sampling, decontamination, environmental monitoring, scene management, and resource acquisition and management.

HMRU also provides CBRNE/hazardous materials (HAZMAT) training to include CBRNE/HAZMAT first-responder programs; CBRNE/HAZMAT technician programs; federally funded training programs; and custom-designed responder programs for hospitals, medical facilities, local and county offices of emergency management services, and law enforcement agencies.

In 1987 the Hazardous Materials Emergency Response Planning (HMERP) Unit was designated through a federal grant to develop and establish a training program to address Occupational Safety and Health Administration (OSHA) regulations regarding training and response for New Jersey's first-responder community. In early 1988 the HMERP unit designated a technical training committee to formulate the training plan and develop the necessary and required components for presentation to the various first-responder agencies. This committee comprised representatives of all emergency response disciplines, including NJSP, local police departments, fire, EMS, HAZMAT, and other relevant state and county government agencies. Through these endeavors, the HAZMAT training program is the leading response training program throughout the nation. The HMERP Unit was responsible for more responders being trained than any other similar programs in the United States. In 1999 the unit was redesignated as the Domestic Preparedness/Hazardous Materials Emergency Response Planning (DPHMERP) Unit. Its primary mission continues to focus on training and planning assistance to response agencies, with additional duties related to emerging domestic preparedness issues.

In response to the World Trade Center attack, anthrax threats, and the other emerging threats, the NJSP in February 2004 created the Homeland Security Branch and Special Operations Section. DPHMERP was again redesignated as the Hazardous Materials Response Unit (HMRU), and along with its existing training, planning and assistance duties, the unit was tasked with operational capabilities.

The training programs provided by the NJSP HMRU address the requirements for individuals who will be responding to hazardous materials incidents. The NJSP HMRU has provided a tiered training curriculum that coincides with the requirements established under OSHA 29 CFR Part 1910.120 (including nonmandatory Appendix E), the National Fire Protection Association (NFPA) 471 standard, the revised NFPA 472 standard, and the U.S. Department of Transportation National Curriculum. The NJSP HMRU has also incorporated national firefighter standards developed under the National Fire Protection Association. The New Jersey Right-to-Know training has been added to reduce training repetition. All HAZMAT courses have also been updated to include modules on terrorism, weapons of mass destruction, and CBRNE response.

In field analytical and response assessment capabilities, HMRU has several pieces of technology to analyze and evaluate unknown materials in the field and transmit data collected to other stations for reach-back capability and further investigatory and confirmatory determination. HMRU has the ability to sustain operations for a protracted period as required to provide support and assistance to NJSP assets and local agencies.

Arson/Bomb Unit

The Arson/Bomb Unit provides the state of New Jersey with statewide fire and explosive investigative capabilities. Bomb technicians also assist members of the Environmental Protection Agency (EPA), the State of New Jersey Department of Environmental Protection (DEP), and the U.S. Drug Enforcement Agency (DEA) with technical assistance with clandestine laboratories and both reactive and unstable chemicals. The unit's personnel are all trained and certified as hazardous devices technicians, hazardous materials technicians, and fire investigators. To be trained and certified, unit members are required to attend lengthy and demanding schooling conducted by the Federal Bureau of Investigation (FBI); the Bureau of Alcohol, Tobacco and Firearms (ATF); and the U.S. military. Advanced training is also conducted outside of the United States in England and Israel. Unit members are responsible for initiating and coordinating both fire and explosives investigations, conducting postblast scene examinations, rendering safe explosive and chemical devices, and destroying unstable commercial explosives, as well as providing V.I.P. bomb sweeps for the U.S. Secret Service, the State Police Executive Protection Unit, and the U.S. Department of State. The unit conducts both fire and explosive lectures for federal, county, and municipal police as well as for members of the fire service. Unit members also assist other law enforcement agencies with special needs in the detection and investigation of explosives incidents or acts of terrorism. Unit personnel have also provided expert testimony in criminal and civil court proceedings in both fire- and explosives-related investigations. In addition to the normal investigative operations, the unit participates in extensive training throughout the state in conjunction with the International Association of Arson Investigators, International Association of Bomb Technicians, FBI, ATF, the Division of Criminal Justice, DEA, and the U.S. Department of State.

The unit's readiness is based on a rapid response with 13 fully equipped SUVs; 3 fully equipped bomb response vehicles strategically located in northern, central, and southern New Jersey; 9 robots; fiber optics and portable X-ray capabilities; large vehicle and personnel-borne improvised explosive device countermeasure defenses; and on-scene explosive materials testing. In addition, the unit is equipped with a mobile explosive containment chamber as well as a weapons-of-mass-destruction containment chamber in preparation for a chemical or biological attack. The unit is also the coordinator of the New Jersey Render

Safe Task Force, which has the capabilities to activate for deployment 52 certified hazardous devices technicians.

Arson/Bomb Unit members have been utilized and have rendered assistance to various government agencies and to a myriad of missions, including the arrest of Yu Kikumura (a member of the Japanese Red Army), the 1993 World Trade Center bombing, the TWA plane crash investigation off the coast of Long Island, the 1997 FedEx plane crash at Newark International Airport, and the September 11, 2001, attacks, as well as assisting with the recovery of human remains in Staten Island. They also provided onsite technical assistance with advanced explosives equipment during the 1996 Summer Olympics in Atlanta, Georgia, and the 2000 World Bank Conference in Washington, D.C. Unit members assisted various federal agencies with field testing and evidence recovery during the 2001 anthrax investigation. The unit has also aided the New Jersey Department of Environmental Protection and Energy, Division of Fish, Game, and Wildlife, by explosively sinking ships off the coast of New Jersey for the man-made artificial reef program. The unit maintains and operates some of the best bomb equipment in the United States and is therefore recognized for its premier abilities and accomplishments.

Technical Emergency and Mission Specialists (TEAMS)

The TEAMS Unit was established in 1978 as a full-time emergency response unit prepared to handle extraordinary police emergencies. During the early years of the unit's existence, members were chosen from personnel of the division's Underwater Recovery Unit, as they had distinguished themselves as highly disciplined and self-motivated troopers. Today the selection process is the most demanding among any in the division. Members are selected based on a written résumé, physical agility test, background investigation, oral interview, and successful completion of the physically and mentally demanding underwater recovery course. The TEAMS Unit comprises three 16-member squads, regionally located in the northern, central, and southern portions of the state.

The TEAMS Unit is a multifaceted entity that consistently meets the challenges of emergency preparedness in a post-September 11, 2001, era. The TEAMS Unit maintains an all-threats–all-hazards–all-crimes methodology toward prevention, protection, response, and recovery. The unit is adaptive and is utilized by federal, state, and local agencies for an array of missions. It is fully operational and can respond to any incident because of the full-time call-out status of its members. The TEAMS Unit maintains an extreme training regimen that is focused on the many disciplines it possesses.

Special weapons and tactics (commonly known as SWAT) is a primary mission for the TEAMS Unit. The unit is utilized to execute tactical intervention strategies for hostage and barricaded-gunmen situations and is also called upon by federal and local agencies to conduct high-risk warrant service entries for some of the state's most dangerous criminal elements. The unit is equipped with technologically advanced systems utilized by SWAT and military units throughout the world.

The TEAMS Unit is also called upon to perform high-angle and confined-space rescue operations during the response and recovery phase of any critical incident. The unit is a support element for NJTF-1. In addition, it has helicopter rappel capabilities either to facilitate a rescue operation or to perform a linear tactical assault.

The TEAMS Unit is utilized to conduct underwater search-and-recovery operations for criminal evidence and unfortunate drowning victims. The unit can perform self-contained underwater breathing apparatus (SCUBA) operations anywhere in the state under some of the most severe maritime conditions. It also performs tactical maritime operations with our Marine Bureau assets.

In counterterrorism operations, the TEAMS Unit is called upon to conduct preventive and protective operations for the state's identified critical infrastructure. The TEAMS Unit is the tactical element for the Target Hardening Response and Emergency Activation Team (THREAT). Because of the TEAMS Unit's tactical capabilities, it can be deployed from the air, sea, or land for any target-hardening mission.

Canine Unit

The NJSP Canine Unit was established in 1987 within the Investigations Section and reassigned to the Special Operations Section within HSB on February 28, 2004. The unit currently consists of 24 troopers assigned cross-trained canine partners who work as a team. The dogs are scent trained to detect the odor of explosives, controlled dangerous substances, and cadavers. The canine teams are cross-trained for patrol functions, which include conducting tracks, evidence or article searches, urban search and rescue, and criminal apprehension. They are available to respond to any request 24 hours a day, 7 days a week.

The primary function of the Canine Unit is to assist federal, state, county, and local law enforcement agencies whenever the services of a police canine are required. The unit assists division personnel in any investigation or motor vehicle stop that necessitates the use of a canine. The canine teams are used to assist in establishing probable cause based on the canine's positive indication for narcotics or explosives. They also assist field operations by utilizing police canines to locate fleeing suspects, missing people, illicit narcotics, explosive materials, and cadaver remains. The unit maintains security enhancement for the Statehouse Complex through high-visibility patrols, explosives sweeps, and covert operations. The unit currently has detachments on the New Jersey Turnpike and at the Atlantic City Airport.

The unit maintains a full-time Canine Training Academy that conducts bi-annual classes in scent and patrol. The trainers provide a 16-week patrol class and a 12-week scent class. They ensure compliance with the Attorney General's K9 Training Standards and utilize the certification methods set forth by the U.S. Police Canine Association. They also conduct monthly in-service training. The Canine Training Academy is located at the former Fort Dix Station, located in Burlington County.

Since the inception of the NJSP Canine Training Academy, more than 260 canine teams from various agencies have successfully completed the training and are certified in scent detection and patrol. The academy also provides in-service training in scent and patrol functions to an additional 70 teams during the ensuing months.

The Canine Unit was selected by the Attorney General's Office to oversee the New Jersey Detect and Render Safe Canine Program. This program includes certification training, maintenance and in-service training, and statewide deployments of the Detect and Render Safe police canines for homeland security and to provide infrastructure security checks. All Canine Unit members conduct police canine education lectures and demonstrations for the Division of State Police; school and Drug Abuse Resistance Education (DARE) presentations; federal, state, and local law enforcement agencies; and the citizenry of the state of New Jersey. The NJSP Canine Unit teams and other canine unit teams throughout the country are a tremendous asset to the law enforcement community.

EMERGENCY MANAGEMENT SECTION (EMS)

The Emergency Management Section supervisor holds the rank of major and also serves as deputy state director of the New Jersey Office of Emergency Management (NJOEM). The section is under the command of the deputy superintendent of homeland security, who is the assistant state director, Office of Emergency Management. The section organizes, directs, staffs, coordinates, and reports the activities of the Communications Bureau, Emergency Preparedness Bureau, and the Recovery Bureau. The supervisor and staff facilitate the flow of information to and from the various bureaus supervised and serve as a conduit for communication with other division entities. The section is also responsible for planning, directing, and coordinating emergency operations within the state that are beyond local control.

The following three bureaus make up the Emergency Management Section:

1. Communications Bureau
2. Emergency Preparedness Bureau
3. Recovery Bureau

Communications Bureau

Radio/Electronics Maintenance Unit (REMU)

The Radio/Electronics Maintenance Unit (REMU) has twenty-three 800-megahertz tower sites with their attendant equipment and buildings, as well as nearly a dozen more low-band tower sites. Its responsibilities include, but are not limited to, all radio and related equipment procurement, distribution, maintenance, and control; maintenance of other public safety equipment; and management of the system's programming. In addition, the REMU provides, maintains, and controls closed-circuit television, tape recordings, and other miscellaneous electronic equipment for the division.

Telecommunications Unit

This unit is responsible for all of the division's telecommunications needs, including telephone service, pagers, cellular phones, and maintenance of the various systems needed to support these services. The unit also handles the liaison between the division and the various service vendors to ensure that the division's needs are met in a timely and proper manner. All of the telephone equipment is owned by the division and is maintained and administered by the Telecommunications Unit. The unit coordinates the 1,000 pagers used by division personnel and is also responsible for funding and maintaining the pagers provided by the Communications Bureau.

In addition to administrating the division's telecommunications needs, the unit also provides security to prevent unauthorized use of the telephone lines. This security is provided by monitoring outgoing calls from division headquarters utilizing the Station Messaging Detail Record and utilizing a barrier code to prevent unauthorized use of the division's toll-free number.

Emergency Preparedness Bureau

The Emergency Preparedness Bureau includes the following units:

- North Regional Unit
- Central Regional Unit
- South Regional Unit
- Radiological Emergency Response Planning and Technical Unit
- Exercise Unit
- Urban Search and Rescue
  - NJTF-1
  - Federal Surplus Property Program

North Regional Unit

The North Regional Unit coordinates emergency management activities throughout the northern 7 counties and 206 municipalities. Of these 206 political subdivisions, 31 receive Emergency Management Assistance (EMA) funding. These EMA-funded jurisdictions result in a unique relationship with regional personnel, who interact with them on a regular basis to ensure their compliance with the EMA work plan. This plan describes the content of the agreement by which they are funded. Regional personnel meet with and evaluate all EMA-funded jurisdictions for year-end reports, development and review of Emergency Operation Plans (EOPs), exercises, and performance review of semiannual and final EMA claim forms.

Central Regional Unit

The Central Regional Unit coordinates emergency management activities throughout the central 7 counties and 192 municipalities, 35 of which receive EMA funding as described above.

South Regional Unit

The South Regional Unit coordinates emergency management activities throughout the southern 7 counties and 167 municipalities, 24 of which receive EMA funding as described above.

EMS regional personnel represent the governor and state director of emergency management at all emergency and disaster situations in the state. They monitor these situations and assure proper response and recovery activities. Response to an incident provides interaction between local and state government that expedites and centralizes the state's response. These activities include state, county, and municipal emergency operations center (EOC) activations, participation in actual operations, and technical assistance during the response and recovery phase. They are also responsible for providing status reports of events, through the appropriate channels, to the Office of the Governor and the attorney general. In the postemergency phase they are responsible for the incident's evaluation and critique. It is also the responsibility of the regional units to assist with the development, review, and compliance of all county and municipal EOPs. The state has currently achieved a compliance rate of 95 percent approved EOPs.

Responsibilities of representatives of the regions fall into 17 functional categories. They are required to attend county, municipal, and other agency meetings; attend and conduct NJSP and NJOEM training; conduct exercises and participate in local exercises; respond to all major incidents and disasters; provide direct EOP development assistance; and conduct compliance surveys.

Radiological Emergency Response Planning and Technical Unit

The Radiological Emergency Response Planning and Technical (RERP&T) Unit develops radiological emergency response plans and procedures for protecting the population in areas within 10 miles of the nuclear power plants located in New Jersey. They develop and conduct an annual exercise of the radiological emergency response plan for each area surrounding one of these plants. They also coordinate the interaction of the state, county, municipal, and federal governments in preparing for response activities to incidents at a nuclear power plant. They also serve as technical advisor to the state director of emergency management and the Governor's Office in incidents or exercises involving radiological materials in transit or at fixed facilities. The RERP&T Unit is responsible for the tasks identified in the Radiation Accident Response Act, 26:2D-37. These tasks center on the development, maintenance, and exercise of the state's radiological emergency response plan. This is also a requirement of the federal government according to 44 CFR 350, Review and Approval of Radiological Emergency Response Plans. The unit develops, coordinates, conducts, and evaluates annual exercises of the plans for the Salem/Hope Creek Plant and the Oyster Creek Plant. These exercises test each major component of the plans and serve to measure the adequacy of those plans and the skills of the responders. Exercises are required annually under the Radiation Accident Response Act. The Federal Emergency Management Agency (FEMA) requires the state to conduct these exercises biennially.

The unit reviews and revises the RERP plans based upon exercise evaluations, revised federal guidance, and changing demographics. The unit also develops, maintains, and exercises standard operating procedures (SOPs), which provide guidance on specific tasks identified in the plans. It develops, provides, and evaluates training to emergency responders on the specific tasks and SOPs identified in the plans. The unit is responsible for the element of the Radiation Accident Response Act that allows for the purchase of equipment necessary for state, county, and municipal governments to implement the missions assigned to them in the plans.

The unit is responsible for supporting the New Jersey Office of Emergency Management's RERP function by calibration, certification, retrofit, repair, stockpiling, and inventory control of radiation protection equipment. This unit's personnel respond on a 24-hour basis to all radiological incidents or potential incidents that occur in or threaten New Jersey. The unit maintains an NRC-licensed radiological protection program for all licensable radioactive material in the custody of the NJOEM. It conducts multilevel radiological training for state, county, municipal, and private-sector personnel and supervises and monitors all radiological exposures to state police personnel.

Exercise Unit

The Exercise Unit was established by the Domestic Security Preparedness Task Force and is composed of representatives from the State Police Emergency Management Section, the Department of Military and Veteran Affairs, and the Division of Fire Safety. Its first priority is the development and delivery of domestic preparedness exercises for all levels of government and the private sector. The unit is also responsible for sponsoring and supporting "all-hazards" exercises for government, private business, and industry throughout the state. The second priority of the unit is the coordination and production of after-action evaluation reports and improvement action plans following the conclusion of exercises, particularly exercises that are funded through the state's federal homeland security grant. When available, the unit also utilizes Emergency Management Performance Grant (EMPG) funding from FEMA to provide financial assistance to opportunistic local jurisdictions and agencies or disciplines seeking to make corrective modifications to their systems or programs following these exercises. The third priority of this unit is the management of exercise data using federally sponsored databases. The unit also maintains the administrative obligation for maintaining program currency with federal and state requirements for incident management and exercise methodology. The unit also collaborates with the Field Training Unit in delivering exercise design and evaluation courses for the expanding emergency management community.

Urban Search and Rescue

The Urban Search and Rescue (US&R) Unit coordinates and completes the administrative requirements for NJTF-1, including personnel and training database management. This unit is able to coordinate the efforts to keep NJTF-1 at a state of readiness commensurate to the FEMA National Urban Search and Rescue (US&R) Response System. They also develop, deliver, and coordinate urban search-and-rescue-related training focused on such topics as structural collapse operations, trench collapse operations, confined space operations, rope rescue (basic and advanced), and swift-water rescue. This unit is also able to run training programs to ensure that they meet and do not exceed the requirements that are set forth by the FEMA US&R System.

NJTF-1 provides advanced search and rescue capabilities to victims trapped or entombed in structurally collapsed buildings. Specially trained in advanced-level search-and-rescue capabilities, NJTF-1 members provide efficient and effective rescue technologies in a planned and measured response system that mirrors FEMA and National Fire Protection Association standards and guidelines. The members of NJTF-1 conduct all search-and-rescue operations in a professional, ethical, and understanding manner to protect the dignity of victims and the local response communities that it serves. Task force members maintain their skills and abilities in technical rescue training with the goal of total preparation for any incidents that may occur now or in the future that require deployments to natural or manmade disasters, hurricanes, floods, conflagrations, explosions, earthquakes, or weapons of mass destruction incidents that are beyond the capability of local emergency services.

Recovery Bureau

The Recovery Bureau includes the following units:

- Public Assistance Unit
- Preparedness Unit
- Mitigation Unit
- Field Training Unit
- Support Services Unit

Public Assistance Unit

The Public Assistance Unit is responsible for managing the Public Assistance Grant Program before, during, and after presidentially declared disasters or emergencies. During a declared disaster, the state of New Jersey, in conjunction with FEMA, provides supplemental aid to communities to help them recover from the effects of a disaster as quickly as possible.

The Public Assistance Unit serves as the principal point of contact for the state. As such, it is responsible for conducting preliminary damage assessments to determine the impact and magnitude of damage and the resulting unmet needs of individuals, businesses, the public sector, and the community as a whole. In the aftermath of a disaster, unit personnel are assigned to FEMA–state preliminary damage assessment teams and coordinate the county and municipal damage assessment efforts as well. The results of damage assessment surveys are assembled by the Public Assistance Unit and presented in a written report for the governor's consideration.

If federal intervention is requested and approved, the Public Assistance Unit provides information about various federal disaster reimbursement opportunities to officials of all eligible state, county, and municipal agencies, as well as designated private nonprofit organizations. The Public Assistance Unit is responsible for coordinating applicants' briefings and kickoff meetings to discuss the parameters of declarations, scope-of-work activities, eligible categories, and documentation required to receive state and federal assistance. The unit also provides technical expertise in the preparation and submission of federal grant and loan applications in accordance with the Robert T. Stafford Act, which requires that eligible assistance be delivered as quickly and efficiently as possible consistent with federal laws and regulations. The unit maintains appropriate files and develops related procedures that comply with all applicable laws, regulations, and Office of Management and Budget circulars governing standard grants management practices. The Public Assistance Unit staff is also responsible for assisting the Field Training Unit in coordinating and delivering training programs and seminars related to the disaster reimbursement process.

Preparedness Unit

Preparedness is an integral part of the disaster management cycle. It is the foundation for reducing losses as well as easing response, recovery, and mitigation efforts. The traditional role of the Preparedness Unit is devising hazard-specific and multihazard plans, educating the public, and conducting public outreach. As our disasters become more complex and our management of emergencies tends towards multihazard integrative efforts, the Preparedness Unit can serve as an outstanding "first line of defense" for combating hazard losses and expenditures. Currently, the unit coordinates a multitude of planning efforts, including maintenance of the state emergency operations plan and the state emergency procedures directory. It also maintains checklists and standardized texts as technical guidance to local government on development of emergency plans and procedures. Hazard-specific plans include those focused on winter storms, hurricanes, reverse-lane evacuations, and drought emergencies.

Significant strides have been made in improving our mental health, special needs, and school planning initiatives as well as establishing a growing public outreach program for natural hazards and evacuation. The unit also maintains liaison and coordination of emergency activities with state departments and various allied support agencies and is responsible for the readiness of the State Emergency Operations Center (EOC). The unit is an integral player in the implementation of "ETeam" technology for use in the state EOC. Another critical role is the timely notification of the emergency management community regarding potentially dangerous weather conditions. The unit administers the National Weather Service's (NWS) StormReady® Program for communities. The unit also coordinates the state's tidal and inland flood warning programs and systems and participates in the New York City Evacuation, Trans-Hudson, and Port Authority emergency planning groups.

Mitigation Unit

The Mitigation Unit has the mission of enhancing state, county, and municipal risk reduction through the development and implementation of mitigation strategies. By definition, hazard mitigation is any sustained action that prevents or reduces the loss of property or human life from recurring hazards. The Mitigation Unit accomplishes this task by implementing and administering several grant-based programs in conjunction with FEMA.

The primary programs administered by the Mitigation Unit are Flood Mitigation Assistance (FMA), Pre-Disaster Mitigation and Pre-Disaster Mitigation Competitive (PDM and PDM-C), and the Hazard Mitigation Grant Program (HMGP). Counties and municipalities are made aware of these programs through letters announcing upcoming grants for which eligible communities in their jurisdictions may apply. Additional workshops are held to further explain available programs, and municipalities are encouraged to apply for grant funds. Upon receiving completed applications, NJOEM will narrow the list of prospective applicants based on existing plans and potential project needs. Follow-up is conducted through extensive use of e-mail communications and phone contact. The state Hazard Mitigation Team will convene to review all applications for funding. Approved project applications and planning grant information are forwarded to FEMA for review and approval.

Upon notification of approval from FEMA, members of the Mitigation Unit notify appropriate municipalities of the award. NJOEM personnel conduct workshops and participate in public meetings with the goal of helping municipalities complete the grant process successfully. Additional workshops are held around the state with presentations given to explain the various programs and their benefits to potential participants. Program partnerships with resource agencies such as the League of Municipalities and professional, civic, and trade-based organizations are utilized to disseminate information and garner public input and inquiries. During a postdisaster period, Mitigation Unit personnel will work closely with all involved communities to assist with the Hazard Mitigation Grant Program in the same manner as is done with predisaster programs.

Field Training Unit

The Field Training Unit (FTU) is responsible for conducting emergency management training courses for state, county, municipal, and private-sector personnel who have emergency management responsibilities or work in related fields. These training programs are designed to assist the public and private sectors in their ability to mitigate, plan for, respond to, and recover from the effects of natural and technological emergencies. All training provided is consistent with training initiatives on the federal level.

The unit offers a variety of interrelated courses designed specifically to improve the professional, managerial, and technical skills of people involved in emergency management. These state-of-the-art training programs are designed to achieve a comprehensive and integrated emergency management system that addresses all hazards at the local, county, and state levels.

Nearly 35 different courses are presented in an adult learning format by teams of experienced, dynamic instructors and subject matter experts. The contributions of the instructors, combined with the interaction of the student body, develop each student's emergency management skills and help them to excel in service to their communities. The unit is also actively involved in conducting emergency-management-related presentations at conferences, seminars, and workshops. The FTU reaches nearly 4,000 students per year in various instructional settings.

Through the coordination of the FTU, the NJOEM has become a participant in the American Council on Education (ACE) College Credit Recommendation Service (CREDIT). To date, approximately one-third of the courses offered by the NJOEM have been recommended for college credit through ACE.

The FTU is also responsible for development of the State Community Relations (CR) Plan. The CR Plan is implemented following a large-scale emergency or disaster. Working in conjunction with FEMA, NJOEM Community Relations officers work door-to-door in disaster areas to collect and disseminate information to and from affected communities; locate individuals who may need special assistance or encouragement to initiate the disaster assistance application process; and identify political, social, religious, ethnic, business, and other interest-group leadership for the purpose of developing a team effort in the recovery process.

Support Services Unit

The Support Services Unit coordinates the development of all Citizen Corps Programs (Community Emergency Response Teams, Neighborhood Watch, Volunteers in Police Service, Fire Corps, and Medical Reserve Corps) throughout the state of New Jersey, with a special emphasis on the urban areas of the state. Coordinating interactions with the New Jersey Volunteer Organizations Active in Disasters (VOAD), the Support Services Unit strengthens ties with the New Jersey business community and maintains a liaison with both the national and the state emergency management communities. The unit's training functions include the Emergency Management Assistance Compact (EMAC) system training and A-Team certification to all branches of state, county, and municipal emergency management coordinators. For preparedness, the Support Services Unit has built and maintains a comprehensive resource directory database of all available emergency response assets in New Jersey.

REGIONAL OPERATIONS INTELLIGENCE CENTER (ROIC)

The Regional Operations Intelligence Center (ROIC) is operational and employs a fusion center concept to facilitate the coordination of governmental departments and partners in emergency management to facilitate a concerted response through the deployment of appropriate resources as they are needed. "The ROIC is supercharging this state's ability to respond to all hazards and all threats," said NSJP Superintendent Colonel Rick Fuentes. "From the vantage point of this fusion center, all dangers to New Jersey's safety can be analyzed and appropriate responses directed. The center will improve our ability to react to everything from terrorism to gang violence to natural/manmade disasters and even major health events." The true strength of the ROIC is found in its partnerships. New Jersey's "fusion center" is the state's hub for intelligence and includes input and personnel from agencies such as the FBI, U.S. Department of Homeland Security, and FEMA; regional partners such as the New York City Police Department (NYPD), neighboring state police departments, and numerous state agencies; and from New Jersey's county, municipal, and nongovernmental partners. The ROIC is also the home for the New Jersey Office of Emergency Management and the state's EOC. It serves as the command center for all state-led emergency response operations, such as natural disasters, chemical or nuclear emergencies, or terror alerts. During emergency response missions, the ROIC serves as the gateway for situational information and requests for aid. It allows a coordinated and measured response by matching requests with resources and personnel from federal, state, and local agencies.

The center's cavernous support room will serve as the focal point of multiagency response during times of large crises. At more than two stories high with 8,644 square feet of floor space, the support room is set up with 100 interdependent workstations that can be assigned to any configuration of agencies involved in an event. Each computer can take the assigned agency's input and send it to a 32-foot-wide by 12-foot-high video wall that keeps all partners apprised of the situation. This screen can show many different inputs at one time. Overlooking the support room from the second floor is the executive conference room, from which the governor and top-level decision makers can view the situation and have videoconferences.

Even during calm times, the ROIC's intelligence activities are continually operating. The Watch Operations component provides situational awareness to state leaders and real-time tactical intelligence to the operators in the field. A key element within Watch Operations is the NJSP Call Center, which handles nearly 70 percent of the 911 cellular calls originating from within New Jersey and adjacent states. The information received by this center is pushed to all elements of the ROIC for identification of trends and patterns, as well as operational deployments. The ROIC's analytical capability is the key to intelligence-led policing that the NJSP has championed. Analysts from all partnering agencies collaboratively link bits of data creating "actionable" intelligence that guides tactical maneuvers in real time or creates crime-fighting strategies for the long term. It is through a robust analytical component that law enforcement has the best chance of averting major criminal acts and responding appropriately to rapidly escalating emergencies to prevent them from becoming more serious.

The state of New Jersey has witnessed the creation of the HSB within the NJSP, through the restructuring of its most specialized bureaus and units. The State Police Special Operations Section offers an operational deterrence-and-response capability through tactical and high-visibility strategic missions, while its Emergency Management Section focuses on planning, preparedness, mitigation, and disaster response. The Technical Response Bureau of the Special Operations Section has been designated as the state's principal response component if a terrorist attack occurs. The HSB is improving its abilities to coordinate its organizational resources to prevent, deter, respond, and recover from a terrorist attack. The NJSP is applying new strategies to improve its capabilities for future threats. It is implementing the National Response Plan of the National Incident Management System, which provides organizational concepts and processes for effective, efficient, and collaborative incident management at all levels. The NJSP has entered into a partnership with 14 other state agencies to form the New Jersey Domestic Security Preparedness Task Force. Through this unity and combination of resources, the state of New Jersey has significantly improved its capability to counter terrorism.

This statement from the Hurricane Katrina after-action report best sums up the actions of the NJSP HSB: "Terrorists still plot their evil deeds, and nature's unyielding power will continue. We know with certainty that there will be tragedies in our future. Our obligation is to work to prevent the acts of evil men; reduce America's vulnerability to both the acts of terrorists and the wrath of nature; and prepare ourselves to respond to and recover from the manmade and natural catastrophes that do occur."

REFERENCES

CNN.com. November 2, 2004. Bin Laden: Goal is to bankrupt U.S. Available online at *edition.cnn.com/2004/WORLD/meast/11/01/binladen.tape/index.html*. Accessed May 5, 2008.

Cooper, Anderson. August 15, 2006. The most dangerous two miles in America. Available online at *www.cnn.com/CNN/Programs/anderson.cooper.360/blog/2006/08/most-dangerous-two-miles-in-america.html*.

The Federal Response to Hurricane Katrina: Lessons Learned. February 2006. Available online at *www.whitehouse.gov/reports/katrina-lessons-learned.pdf*.

National Security Task Force on Energy. 2006. Energy Security in the Twenty-First Century: A New National Strategy. Available online at *www.americanprogress.org/kf/energy_security_report.pdf*.

New Jersey Domestic Preparedness Task Force. 2006. 2004/2005 Progress Report. Available online at *www.njhomelandsecurity.gov/dsptf/NJDSPTF-04-05-021706.pdf*.

New Jersey Office of Homeland Security and Preparedness. 2007. The Terrorist Threat to Energy Infrastructure.

New Jersey State Police. Homeland Security Branch. Available online at *www.njsp.org*. Accessed April 26, 2007.

New Jersey State Police. 2006. Practical Guide to Intelligence-Led Policing. Available online at *www.njsp.org/divorg/invest/pdf/njsp_ilpguide_010907.pdf*.

State of New Jersey, Office of the Governor. January 24, 2007. Governor Corzine and law enforcement officials open state-of-the-art emergency management facility. Press release available online at *www.state.nj.us/governor/news/news/approved/20070124a.html*.

United States Department of Homeland Security. 2005. National Preparedness Goal. Homeland Security Presidential Directive 8: National Preparedness. Available online at *www.ojp.usdoj.gov/odp/docs/InterimNationalPreparednessGoal_03-31-05_1.pdf*.

United States Department of Homeland Security. 2006. National Infrastructure Protection Plan. Available online at *www.dhs.gov/xlibrary/assets/NIPP_Plan.pdf*.

15

The Problem of Oil and Natural Gas Pipeline Security*

S. G. Serebryakov,
Russian Academy of Sciences Institute of Oil and Gas Problems

The natural gas produced in Russia is transported through major gas pipelines linked to Russia's Unified Natural Gas Supply System (see Figure 15-1), the largest such system in the world. The total length of all pipelines in the system, which belongs to the Open Joint-Stock Company Gazprom, is 156,300 kilometers. It includes 268 compressor stations with a total of 4,078 pumping units with a 44.8 million kilowatt capacity, as well as 3,818 natural gas distribution stations.

As of December 31, 2005, the average length of service for major natural gas pipelines was 22 years. Their stable operation is ensured thanks to the introduction of progressive methods for diagnostics, scheduled maintenance, and repairs. Gazprom is implementing a comprehensive program for reconstruction and technical upgrading of natural gas transmission lines, compressor stations, and underground storage facilities for the period 2007-2010. The primary goals of the program are to improve the efficiency of the gas transmission system, ensure the transport of planned volumes of gas and the reliable operation of the transmission system, and improve the industrial and environmental safety of all its components.

The Unified System is today operating at full capacity. In 2005, Gazprom

*Translated from the Russian by Kelly Robbins.

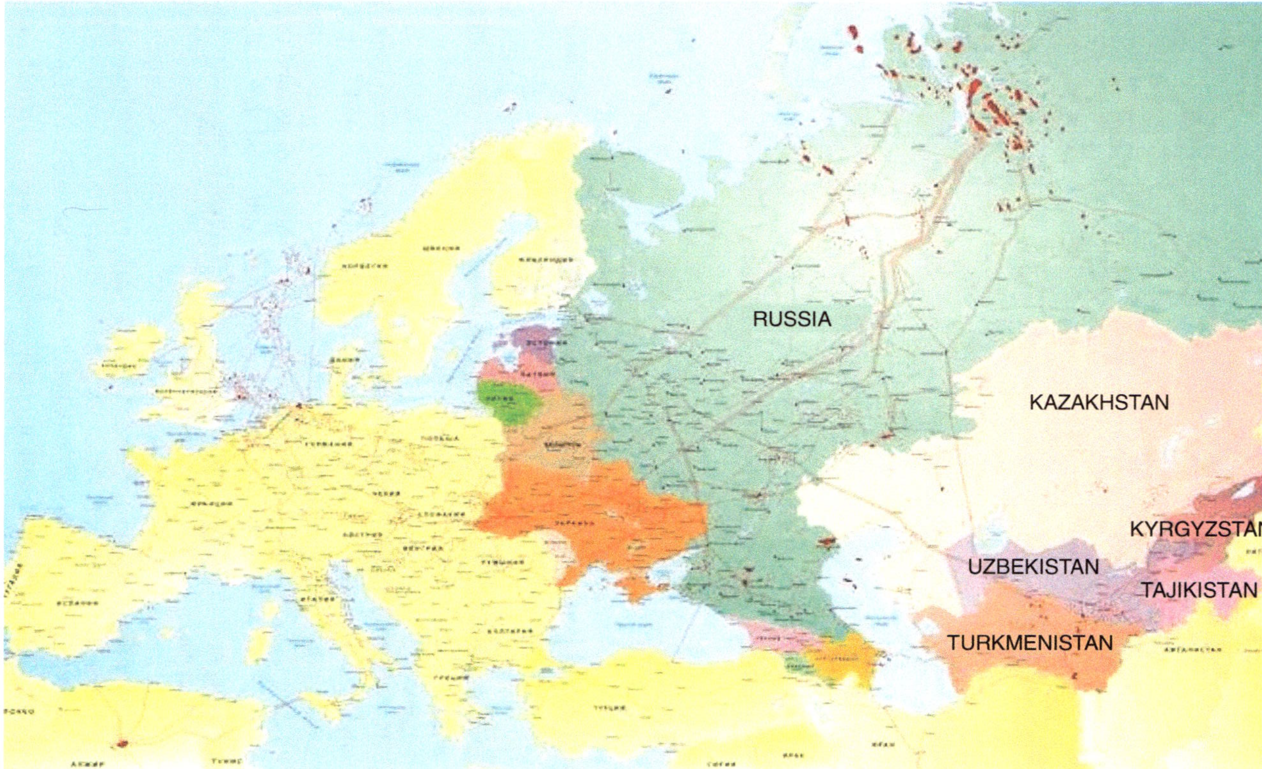

FIGURE 15-1 Russia's Unified Natural Gas Supply System.

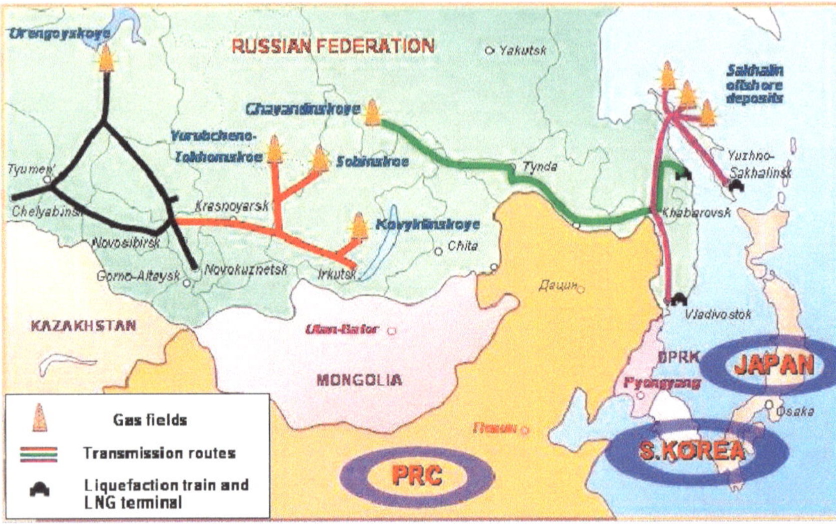

FIGURE 15-2 Planned routes of major natural gas pipelines.

extracted 547.9 billion cubic meters of natural gas. Taking into account independent producers and those from the Central Asian states, the system transported a total of 699.7 billion cubic meters of natural gas. Even today its transmission capacity needs to be increased by 35 billion cubic meters, with further increases necessary in the future, given that by 2020 Gazprom plans to extract 580-590 billion cubic meters of natural gas with up to an additional 170 billion cubic meters from independent producers. The planned routes of the major gas pipelines are shown in Figure 15-2.

The 24 underground natural gas storage facilities located in areas of major gas demand are an essential element of the Unified System. They make it possible to handle seasonal fluctuations in natural gas demand, reduce peak system loads, and ensure flexible and reliable gas transmission. Three underground storage facilities are under construction, including one near Volgograd that will be the largest of its kind in Europe, with a volume of 800 million cubic meters and a daily output capacity of 70 million cubic meters.

With a total length of more than 46,000 kilometers, the Transneft company's unified system of major oil pipelines (see Figure 15-3) transports 99.5 percent of all oil produced in Russia both to refineries and for export to the countries of the Commonwealth of Independent States, Poland, Germany, Slovakia, and Hungary through the Druzhba oil pipeline system and through deep-water oil transfer terminals on the Black and Baltic seas. Transneft serves a territory twice as large as

FIGURE 15-3 Unified system of major oil pipelines of Transneft and nearby foreign countries.

that of the U.S. oil supply system and provides transport services to oil-producing enterprises in the republics of Kazakhstan and Azerbaijan.

At the end of 2005, the total length of Russia's major pipelines was more than 231,000 kilometers, including the following:

- Major natural gas pipelines: 161,100 kilometers
- Major oil pipelines: 49,000 kilometers
- Major refined product pipelines: 19,500 kilometers
- Ammonia pipelines: 1,400 kilometers

All of these facilities present a significant danger to personnel, the public, and the environment.

A report submitted in 2005 by the Russian Federal Service for Environmental, Technological, and Nuclear Oversight (FSETNO) shows that the following factors present the primary threats to the integrity of major pipeline transmission facilities:

- Intensive development of stress corrosion processes on large-diameter major natural gas pipelines due to deterioration of the protective sealant coating the pipelines, which were constructed 15 or more years ago. Whereas from 1991 through 1996, the rate of accidents due to this cause was about one-fourth of all accidents in the Gazprom system; from 1998 through 2003, accidents due to this cause represented one-third of the total; and in 2005 this figure was already more than 50 percent.
- Significant growth in the number of cases of unauthorized connections to oil and petroleum product pipelines with the aim of stealing the products being transported. This increase in thefts has been particularly acute in the republics of Dagestan and Chechnya; Samara, Nizhny Novgorod, and Saratov oblasts; and Stavropol and Krasnodar territories.
- Accidents due to shoddy construction and installation work resulting from the lack of an effective system of technical monitoring of design specification compliance during intensive construction of major pipeline transport facilities in the 1970s and 1980s.

An analysis conducted by FSETNO of the results of investigations of accidents occurring in 2005 is presented in Table 15-1.

Table 15-2 presents FSETNO's analysis of accident and injury statistics for major pipeline transport operations for 2005 compared with the same figures for 2004.

The most significant accidents in 2005 were those at the Petrovsk-Yelets main natural gas pipeline on January 18, 2005, and the Khadyzhensk-Psekupskaya-Krasnodar main oil pipeline on August 7, 2005.

On January 18, 2005, the major natural gas pipeline Petrovsk-Yelets (built

TABLE 15-1 Results of Investigations of Accidents in 2005

Cause	Natural Gas Pipelines	Oil Pipelines	Product Pipelines	Total
1. External mechanical impacts, including	3	12	5	20
Cutting	–	8	1	9
Construction equipment	3	4	4	11
Terrorism	–	–	–	–
2. Corrosion damage	14	–	–	14
3. Shoddy construction or installation work	3	2	-	5
4. Operator error	1	-	1	2
5. Defective parts or materials received from manufacturer	2	2	–	4
Total	19	13	3	45

TABLE 15-2 Accident and Fatal Injury Rates for Major Pipeline Operations in 2004 and 2005

	Number of accidents			Number of fatalities due to injury		
Pipelines	2004	2005	+/-	2004	2005	+/-
Natural gas pipelines	29	19	–10	2	2	0
Oil pipelines	19	13	–6	3	2	–1
Refined product pipelines	0	3	+3	1	–	–1
Total	48	45	–13	6	4	–2
Total length of pipelines (in thousands of km)	231	231	0			

in 1981 and owned by Gazprom and the Mostransgaz Limited Liability Society) suffered damage at its 316-kilometer mark that blew out 55 meters of pipe and caused the gas to ignite. The accident resulted from the formation during pipeline operations of lengthwise cracks in the surface of the pipe, which, at the moment of the accident, failed to provide the expected stability and led to the pipeline segment being destroyed. The economic impact of the accident was 3,710,900 rubles.[1]

On August 7, 2005, the Khadyzhensk-Psekupskaya-Krasnodar main oil pipeline began discharging oil into the Chiby Canal at its 80-kilometer mark in the Republic of Adygeya. The accident was caused by thieves making unauthorized access to the pipeline to steal oil. The costs of dealing with the accident and its consequences totaled 3,732,185.54 rubles.[2]

To ensure the industrial safety of the major pipeline transport facilities of Gazprom, Transneft, and Transnefteprodukt, the Comprehensive Programs for Facility Diagnostics, Technical Upgrades, Reconstruction, and Major Repairs have been developed and coordinated with FSETNO and are currently being implemented.

It must be noted that protecting pipelines from terrorism has taken on increasing significance in recent years, particularly after September 11, 2001.

The results of FSETNO inspections in 2005 of the level to which hazardous production facilities are protected against terrorist acts showed that on the whole, all enterprises with such facilities have developed a system of measures to prevent terrorist acts and have made agreements with specialized services to protect them. The grounds of the most important facilities are surrounded by fences or other protective barriers. Meanwhile, many facilities (wells, pipelines, and so forth) are unprotected; therefore, measures are in place for them to be patrolled regularly. All facilities are equipped with telephone hotlines directly connected to emergency services and security dispatch centers. Plans for new hazardous facilities include the installation of external video observation centers.

Individual terrorist acts have been carried out against pipelines in Russia, primarily during the period of military actions in Chechnya. On April 15, 1996, a bombing severed a major natural gas pipeline 1,200 millimeters in diameter on the left bank of the Terek River in Shelkovskaya Region in the Republic of Chechnya. When the explosive device was detonated near where the gas pipeline emerges from underground, the entire pipeline was blown off its supports. The blast produced a crater 33 by 29 meters in area and 10 meters deep. The pipeline break led to a fire that burned a total of 25,000 square meters on both banks of the Terek River.

On June 14, 1999, sabotage caused an accident at the 124-kilometer mark on the linear portion of the Grozny-Baku main oil pipeline, which is owned by the Open Joint-Stock Company Chernomortransneft and the Stock Company Transneft. Placed in service in 1983, the Grozny-Baku oil pipeline has an operating pressure of 4.3 megapascals and transports a mixture of Azerbaijani and Dagestani oil. The raised segment of the pipeline at its 124-kilometer mark in the Yaryk-Su River bed is suspended on supports and constructed of pipe that is 720 millimeters in diameter. The accident was caused by the energy effects of an explosive device placed under the pipeline. A total of 199 cubic meters of oil (169 metric tons) was spilled, contaminating about 3 hectares of nearby territory in the river basin. The costs of dealing with the accident and its consequences totaled 664,896 rubles.[3] These examples provide a graphic demonstration of the damages caused by terrorist attacks.

The potential terrorist threat and the increased number of cuts in oil pipelines for the purpose of unauthorized removal of oil have required Transneft to take certain measures to prevent damages to elements of the energy infrastructure. In 2000 the company began working to create a concept for an effective vertically integrated corporate security system. It created the Security Systems Department, including mobile armed groups and economic security subunits for the company as a whole and for its subsidiaries, to provide physical protection for oil transport facilities. To prevent pipeline damage and oil thefts due to unauthorized pipeline cuts, the Security Systems Department is introducing several technical capabilities and closely tracking all modern research developments in this field both in Russia and abroad. In recent years, security equipment has been installed at 78 percent of the company's facilities.

Pipelines present a convenient target for terrorists, inasmuch as a simple explosive device can knock them out of commission for weeks. It is for this reason that they have become the focus of sabotage in Iraq.

According to information from the Institute for the Analysis of Global Security (United States), since the end of the war declared by George Bush in April 2003, 37 attacks on pipelines, oil facilities, and their personnel were recorded in 2003, 147 in 2004, 100 in 2005, 100 in 2006, and 5 in January 2007. Most of these attacks occurred on pipelines leading to Turkish and Syrian terminals on the Mediterranean Sea, at the Bayji refinery complex 200 kilometers north of Baghdad, and at oil facilities south of Basra, where more than two-thirds of Iraq's oil is extracted (see Figure 15-4).

Iraq's proven oil reserves total 15.5 billion metric tons (9.6 percent of world reserves), ranking the country fourth behind Saudi Arabia, Russia, and Iran. Meanwhile, after falling to 65.7 million metric tons in 2003 and rising to 99.2 million metric tons in 2004, oil production again fell by 10 percent to 89.5 million metric tons in 2005. Iraqi oil could take the pressure off world markets in the face of high demand by China, the problems with Iran's nuclear program, and unrest in Nigeria's oil-producing region in the Niger Delta. However, the country is not even meeting its own domestic needs. As a result, Iraq imports refined petroleum products at very high prices at the same time that it could be increasing its own oil exports and earning money to restore its economy.

"Every day that oil shipments are paralyzed costs us $60 million," Iraqi oil minister Tamir Gadban has declared. All of this has caused Iraq to lose more than $10 billion from oil sales, undermined prospects for the country's reconstruction, and led to a situation in which oil companies are not taking the risk of investing in the development of the Iraqi oil and gas industry.

The success of terrorist acts in Iraq has led terrorists in other oil-producing countries to turn their attention to pipelines and other oil industry facilities. In December 2004, insurgents attacked an oil field in Sudan. In India, separatists claimed responsibility for several attacks on oil pipelines in the state of Assam, the source of about 15 percent of India's oil output. Rising demand for oil in the country makes its economy increasingly sensitive to supply disruptions. In Turkey, Kurdish partisans carried out a series of strikes against oil pipelines. At-

FIGURE 15-4 Map of Iraq's oil pipelines.

tacks on oil-drilling platforms in Nigeria in 2006 led to a halt in oil production in that country.

However, the greatest disruptions in oil supplies to the world market would be caused by actions against the oil pipelines of Saudi Arabia, which produces about 25 percent of the world's oil and which has about 17,000 kilometers of pipelines throughout the country, primarily located underground.

Economists have calculated that the risk of terrorist attacks has already caused the price of oil to rise by about $10 per barrel as a sort of insurance premium. Terrorists clearly understand that oil price increases as a result of sabotage against oil and natural gas pipelines are felt very keenly not only by the U.S. economy, which lost about $40 billion for this reason in 2004, but also by the world economy.

The most obvious way to provide increased security for pipelines is to establish security patrols and create buffer zones along their routes in which unauthorized access is prohibited. In Iraq an entire army of 14,000 guards has been deployed along pipelines and at oil wells and refineries.

Systems for detecting irregularities and complex modern systems for monitoring particularly vulnerable points could play an important role in protecting pipelines. These systems, which use supersensitive seismic monitoring devices, could provide early warnings if saboteurs were to approach a protected area. Such remote monitoring systems for the pipeline network could be very expensive; however, they would make it possible to avoid the costs of supporting a significant contingent of troops to protect the network, as personnel needs would be limited to small rapid-response groups.

These systems could be augmented by observation from the air, including using pilotless drones capable of flying for up to 30 hours at medium and low altitudes and transmitting high-resolution images to a central station for subsequent processing. There have been reports of the development of pilotless aircraft equipped with automatic weapons, which could be used against saboteurs. Unfortunately, the majority of countries where such systems would be most effective lack the necessary financial resources to acquire them. In such cases, even fences and walls could be used as protective measures to prevent access to facilities. New pipelines must be laid underground. This increases their construction costs, but the return on investment is rapid. It is also important to reduce the time between pipeline damage and repair; the shorter the time, the less damage is done. With this in mind, it would make sense to reduce the length of damaged pipeline segments that must be repaired.

However, it must be understood that ensuring the security of natural gas and oil pipelines is a rather complex problem, the resolution of which is determined by improved equipment and technology for new pipeline construction, more reliable diagnostics, modern means of rapidly eliminating the consequences of accidents, and, on the other hand, development of effective measures and equipment for preventing terrorist attacks against elements of the oil and gas infrastructure. No matter what new equipment or capability may be proposed, it will only increase the cost of a barrel of oil, which has already reached a colossal level. As long as oil and natural gas are the foundation on which the world economy functions, the threat of such attacks will obviously remain, and new achievements in the sphere of their prevention will inevitably increase the price of a barrel of oil.

NOTES

1. Approximately $132,343 at the exchange rate prevailing at that time.
2. Approximately $131,323 at the exchange rate prevailing at that time.
3. Approximately $26,810 at the exchange rate prevailing at that time.

16

U.S.-Russian Collaboration in Combating Radiological Terrorism

John F. Ahearne, Sigma Xi, The Scientific Research Society

Later in 2007, the U.S. National Academies will publish the report of the Committee on Opportunities for U.S.-Russian Collaboration in Combating Radiological Terrorism.[1] The following are key extracts from this report.

Packaging conventional explosives with radioactive material and detonating a radiological dispersal device (RDD) to kill and terrorize people—the "dirty bomb" scenario—is, unfortunately, readily within the means of some terrorist groups. The International Atomic Energy Agency (IAEA) reports that radioactive materials needed to build an RDD can be found in almost any country in the world and that more than 100 countries may have inadequate control and monitoring programs necessary to prevent or even detect the theft of these materials. The agency also reports numerous incidents of illicit trafficking in radioactive materials, including ionizing radiation sources (IRSs) used in medical, agricultural, and industrial applications. Potential links of such trafficking with international criminal organizations heighten the concern about these materials falling into the hands of terrorists, who could use them in RDDs or in other ways to threaten populations.

The challenges in preventing detonations of RDDs are immense, and they will persist for many years. Hundreds and perhaps thousands of inadequately protected IRSs that are considered dangerous by safety standards adopted by the IAEA are present in many countries. Some are in use, some are in storage, and some are awaiting permanent disposal. Also, some IRSs have been simply abandoned by their legal custodians, since there were no financially affordable disposal pathways for those that had exceeded their useful lifetimes or were no longer needed. Poorly protected IRSs, and particularly those that have been abandoned, can become easy prey for terrorist groups.

The IAEA is leading international efforts to enhance security of IRSs. The agency has prepared the Code of Conduct on the Safety and Security of Radioactive Sources and supporting documents that provide guidance for ensuring both the safety and the security of IRSs. Also, it has long had a technical assistance program to help member states improve the security of IRSs. The U.S. Department of Energy (DOE), in close cooperation with the IAEA, has undertaken a limited but important set of cooperative activities with other countries in enhancing security of IRSs in those countries. Programs in Russia have been an important component of this global effort.

The committee decided to concentrate its efforts on the radiological terrorism threat posed by inadequately protected IRSs in Russia and on feasible approaches to upgrading the security of IRSs in Russia. Based on site visits by committee members, consultations with dozens of Russian and U.S. specialists, and reports prepared by our Russian collaborators, the committee concludes that shortcomings in the security and life-cycle management of IRSs in Russia present a serious problem. Hence, the special attention directed to security of IRSs in Russia within DOE's global programs is very appropriate.

A successful RDD detonation in Russia, or indeed in any country, poses serious problems for the United States. Such attacks could provide a "proof of principle" for terrorists who have not yet used radiological weapons, possibly encouraging copycat attacks by terrorists in the United States or against U.S. interests abroad. An RDD attack in Russia or elsewhere could undermine the credibility of the IAEA as an effective international organization for ensuring nuclear safety and security, just at a time when the United States is firmly committed to strengthening this organization to deal with nuclear security and nonproliferation issues worldwide. The United States has considerable interest in helping to ensure that the security of IRSs in Russia meets an international level of acceptability and that Russia improves the full life-cycle management of its IRSs.

The committee is deeply concerned over the continuing decline in the level of DOE resources being allocated to the cooperative program in Russia. DOE should move forward promptly to work with Russian counterparts to address the most urgent problems and help them develop and implement their program. Of special relevance to development of a comprehensive Russian program for addressing the security of IRSs is the approach of the Federal Atomic Energy Agency (Rosatom) in the area of "safety" of IRSs and radioactive waste. Rosatom has developed and regularly articulates a comprehensive overview of safety-related actions that are needed and are under way. According to Rosatom officials, this overview is very helpful in guiding the national effort.

During the past several years, and particularly since September 11, 2001, international concern over the use by terrorists of radioactive material as a radiological weapon has increased considerably. The possibility of the detonation of an RDD, often referred to as a dirty bomb, which has radioactive material packed in or around conventional explosives, has been the focus of much of this apprehension. Press reports of illicit trafficking in radioactive material, Web chat attributed to terrorist groups, and discovery of primitive drawings of dirty bombs in the possession of international terrorist groups have heightened the concern.

In addition to misuse of radiological sources considered in this study, radiological terrorism could be carried out by sabotage of a nuclear facility, waste site, or transport container. Terrorists might attempt to detonate, set fire to, or otherwise cause serious dispersion of radioactive material located within the target area. However, this report focuses primarily on dispersion of radioactive material. Terrorists might acquire by theft or other means nonfissile radioactive material and disperse such material with conventional explosives in an RDD. Other forms of radiological terrorism include the dispersion of radioactive material through public pathways, such as water supplies, roadways, or indoor heating or ventilation ducts. Another form of radiological terrorism is posed by radiological exposure devices, which are radiation sources placed in public places that simply irradiate nearby persons rather than dispersing the radioactive material. Funding and time limitations led the committee to concentrate on RDDs.

Unfortunately, few if any countries have given sufficient attention to the security of IRSs during their entire life cycle (from fabrication to final disposal), particularly after they have exceeded their useful lifetime or are no longer needed. In recognition of the security importance of ensuring that unwanted IRSs are not left unattended, in 2002, DOE moved its Orphan Source Recovery Program to its threat reduction organization. Similarly, the IAEA greatly expanded its Code of Conduct on the Safety and Security of Radioactive Sources and associated IRS programs to go beyond safety concerns and focus as well on security, including orphan source recovery.

Several Russian research organizations have been analyzing on a broad basis developments in Russia relevant to this study. For example, the Nuclear Safety Institute (IBRAE) of the Russian Academy of Sciences has published many articles on radiological terrorism concerns in Russia and other countries, including security of IRSs. A study commissioned by the National Research Council and carried out by IBRAE provides a Russian perspective on many aspects of the topic of this report. Findings of the study are included in this report as appropriate. The Institute of Chemical Technology of Rosatom, in cooperation with several other Russian institutes, has prepared a series of reports on distribution of radioactive material and radioactive contamination in Russia, under a broadly based program entitled "The Radiation Legacy of the Soviet Union."

Because IRSs have beneficial uses inextricably integrated into medicinal, agricultural, industrial, and research activities, and because their use will increase as the world becomes more industrialized, they cannot simply be locked up or eliminated. The challenge for governments is to expand their efforts to keep IRSs out of the hands of terrorists through life-cycle management while at the same time preparing to manage the consequences if dirty bomb events occur.

Of course, RDDs cannot trigger a nuclear explosion with its familiar mushroom cloud. Unlike nuclear weapons, they cannot instantly kill tens to hundreds of thousands of people and obliterate a city. Thus, the concept of radiological terrorism is quite different from the possible use of nuclear weapons, and linking the two threats can hinder efforts to properly define the risks and to prevent events.

Radioactive material dispersed by an RDD may cause serious radiation health effects for a limited number of exposed people and indeed may result in some deaths. But the gravest consequences of detonation of an RDD are more likely to be the spread of contamination requiring evacuation of large numbers of inhabitants of the affected area; short and long-term economic disruption that could extend well beyond the contaminated area by impacts on transportation, financial, and other sprawling infrastructure systems; incitement of psychological trauma among individuals and groups that are exposed to radiation or believe they have been exposed; and attendant social or political instability.

The possible consequences of an RDD incident can only be predicted through analysis of the impacts of major radiation accidents and other types of relevant events and from hypothetical scenarios. The IBRAE report postulates several scenarios and discusses possible health, economic, and disruption impacts. Of particular concern are cleanup problems associated with different radionuclides.

When properly packaged, adequately shielded, and appropriately handled for their intended use, IRSs are safe, even when they contain the most lethal radionuclides. However, if the shielding is removed and the containers are breached either intentionally or unintentionally, the radioactive material in many IRSs can injure or perhaps even kill exposed persons and could seriously contaminate large areas.

The committee is unaware of any authoritative estimates of the total number of IRSs that are in use or storage throughout the world. Worldwide inventories up to 10 million have been reported. The committee believes that the number is in the millions but cannot be more precise using available data. Countries that have produced and distributed IRSs should attempt to calculate the quantity of radionuclides produced and distributed to date to help establish an upper bound on an overall estimate of inventories. This information would assist in determining the level of resources that should be devoted by governments to combating radiological terrorism.

The IAEA has developed the accepted international standard for categorizing IRSs according to the safety aspects of each type of IRS.[2]

Orphan sources are IRSs that are considered by their legal custodians as no longer needed and have simply been abandoned. They are substantial problems in many countries, including Russia. A particularly worrisome security vulnerability results when IRSs are no longer needed and a clear and affordable disposition path does not exist.

An important concern in the United States is the possible malevolent use of unaccounted-for IRSs, a problem that began to gain increased attention well before September 11, 2001. However, the lack of available disposal pathways for some IRSs leaves licensees limited viable options when IRSs are no longer needed. The regulatory framework is not well prepared to deal with this problem, and large numbers of excess and unwanted IRSs have accumulated in storage with no routes for permanent disposal.

In 2006 the DOE target was to recover and secure an additional 2,000 sources. These efforts clearly lower the probability that radiological material will fall into the hands of terrorists within the United States, and this experience should be instructive in helping to address security weaknesses in Russia and other countries. Achievements within the United States of DOE's Global Threat Reduction Initiative (GTRI) and its predecessor programs since 1997 have included recovery of more than 12,000 high-risk radiological sources.

As of August 2005, DOE had participated in installation of security upgrades and new construction at more than 100 sites in the former Soviet Union. This activity included construction of new, secure storage facilities for IRSs in Uzbekistan, Moldova, Tajikistan, Kyrgyzstan, and Georgia, with a facility under construction in Azerbaijan. Security upgrades have included hardened doors and windows, intrusion detection systems, and response-force equipment. Additionally, the IAEA and the Russian firm Izotop have assisted several countries in dismantling irradiators that are no longer used and in transporting IRSs to secure storage.

RADIATION SOURCES IN RUSSIA

Russia possesses a very large number of IRSs, dating from production during Soviet times and continuing to today with production in Russia. The number of IRSs has been reported by IBRAE to be more than 500,000, but experts from this institute and other organizations readily acknowledge that the number is probably much greater and could be as large as 1 million or more.

Of special concern are the thousands of high-activity IRSs of IAEA Categories 1, 2, and 3 that were produced during the Soviet era and distributed throughout the Soviet Union. A significant number were also exported to other states with close ties to Moscow. Many of these IRSs are still located in other former Soviet states as well as in Russia. A particularly troublesome aspect of the Soviet nuclear legacy is the large number of inadequately protected high-activity IRSs that have been used as radioisotope thermoelectric generators (RTGs) to supply small amounts of electrical power at remote sites, primarily in Russia, with a few also sent to outlying states.

Security of IRSs eroded rapidly during the dramatic political and economic transitions in Russia in the early 1990s. The state system was in turmoil. The institutions that had IRSs in their possession lost much of their financial base, and individuals in charge were often changed with little advance notice. Indeed, the authority vested in various components of the regulatory system was in a state of flux, and the government soon lost track of very large numbers of IRSs. Many privatized institutions stopped reporting their inventories to the government. Some soon declared bankruptcy and simply walked away from their responsibilities for controlling and accounting for IRSs. Often scavengers collected what they thought was usable metal from equipment that may have contained IRSs.

Reports of IRSs being found abandoned in public places and in dormant industrial facilities in recent years have been manyfold. The historical political and economic upheaval has dramatically affected the physical protection, control, and accounting of IRSs. The need to upgrade security is clear.

In general, Rosatom has the ultimate responsibility for control and accounting of IRSs within the country with the exception of sources under the purview of the Ministry of Defense. (The committee did not have adequate information to comment on security of IRSs within the military complex other than observations concerning RTGs.) Organizations that possess IRSs have the primary responsibility for the physical protection of IRSs and for providing information to Rosatom, directly or indirectly, concerning the control and accounting of their inventories of IRSs.

Russian officials admit that enforcement is a problem. When organizations do not comply with Rosatom requirements for providing data on their inventories, Rosatom has two options: (1) send a reprimand to the organization or (2) report the violation to the Federal Service for Environmental, Technological, and Nuclear Oversight (Rostekhnadzor), which has the authority to withdraw the organization's operating license. Rostekhnadzor officials pointed out to the committee that should a license be withdrawn, the agency has no means to remove or secure the IRSs that are affected. Rosatom is attempting to manage problems of inadequate security on a comprehensive basis, but it has limited enforcement authority that is distributed among several organizations and their affiliated branches operating at both the federal and the local levels. As in the United States, a weak link in the regulatory framework is end-of-life management of IRSs. Responsibilities become unclear when IRSs are no longer needed and are abandoned.

On the whole, in Russia many skilled and dedicated people with relevant expertise are working on improving legal and regulatory systems related to IRSs and implementing security programs at the facility level. However, while organizational responsibilities seem to be reasonably well defined, the committee believes that the information presented in this report, including reported efforts by Chechen insurgents to use IRSs for malevolent purposes, calls for greater efforts by the Russian authorities and international partners to upgrade security efforts for IRSs.

The IAEA's Code of Conduct on the Safety and Security of Radioactive Sources calls for security measures to deter, detect, delay, and respond. In considering security enhancements in Russia, the following specific steps might be considered, based on information available to the committee concerning conditions at the facility level:

- Improved personnel and vehicle access checkpoints equipped with appropriate detection devices

- Upgraded perimeter surveillance systems and security alarms
- Routine surveillance at IRS storage locations
- Improved communication and alarm capabilities within facilities, with connections to external response forces
- Power backup supplies and associated lighting systems
- Special secure containers for storage and shipment of IRSs

Although the Russian government has begun to put some of the key building blocks in place, the system of accounting and control should be strengthened. This step is especially important, since accountability for many IRSs was disrupted during the transition from Soviet to Russian control. Also of critical importance are effective procedures for addressing security of the tens of thousands of IRSs in the range of 1 to 100 curies that are in circulation. This is an enormous challenge for Russian organizations. Although steps are under way to improve the accountability of IRSs, many more steps are needed to have adequate life-cycle management.

An aggressive program of disposing of unneeded IRSs could reduce the number of organizations and the locations within organizations that require protection. If no clear and affordable disposition path is available, then some facilities may resort to other means to hide or just abandon sources because they cannot afford to secure them properly or ship them to a disposal facility.

Rosatom has developed a comprehensive approach to providing emergency rescue and related services. A crisis center operates continually within Rosatom both to coordinate information and to manage day-to-day activities. Special emergency services have been identified throughout the country, with essentially all of Rosatom's resources on call should a need arise. Special antiterrorist forces have been organized for deployment from both closed nuclear cities and other cities. Special transportation units are available, and even a special militarized mountain rescue brigade is on call.

Rosatom is but one of many ministries and agencies prepared to respond to an RDD attack. The emergency response ministry (EMERCOM), the health authorities, the police, the security services, and many other federal and local organizations would play important roles. The immediate responsibilities and indeed the longer-term structure of the response would depend to a considerable degree on where the incident occurred and the seriousness of the ensuing contamination. Whereas Moscow appears to have impressive capabilities and experience for responding to an RDD attack, the remainder of the nation's cities are less well prepared. In many cities, the financial difficulties in the 1990s severely weakened staffs and equipment capabilities to respond to any type of crisis. But in Moscow, the committee observed a level of sophistication regarding emergency operations and response capability that should be of considerable interest to the U.S. Department of Homeland Security.

With the passage of time following an incident, mistrust of governmental assessments and decisions will most likely arise among some elements of the population. The Russian government has not been strong on risk communication in the past—a situation that is not unique to Russia. While government services for evacuees are likely to be substantial in scope, as they have been with previous accidents and attacks, the quality and sustainability of such services may not be high. The committee noted one apprehension among some Russian colleagues regarding the effects of a radiological attack that is not voiced in the West, namely, the potential for political instability that an effective RDD event might cause as various elements of the population lose confidence in the government's ability to protect its citizens.

In summary, the security of Russian IRSs has several weak links, often associated with lack of adequate financial resources. Russia was fortunate to progress through the most difficult transition years in the 1990s without a major radiological incident despite serious vulnerabilities. During the past few years, many significant security enhancements have been made, some through the DOE cooperative program. But more work is needed before Russia achieves an internationally acceptable level of security for its inventory of IRSs.

In addition to shoring up the security during all phases of the service life of IRSs, a comprehensive life-cycle management approach is essential, with adequate human resources. At the same time, Russia is currently demonstrating that it can safely and securely manufacture and distribute IRSs worldwide on a competitive basis. In this revenue-generating area, the necessary infrastructure seems to be quite adequate. Russia also has a wealth of nuclear science and technology expertise sufficient to develop, manufacture, and deploy state-of-the-art radioactive material detection equipment for protection of its own borders. This equipment is also competitive in the world marketplace and can be offered to other nations for the protection of radioactive materials.

The United States has a direct and substantial interest in the security of IRSs in Russia. While thefts of IRSs close to U.S. government and U.S. private-sector facilities would be of great concern (for example, Moscow, St. Petersburg, Yekaterinburg), thefts at more distant locations where large amounts of dangerous radionuclides are located should also be of concern. In short, it is difficult to prioritize security upgrades on the basis solely of location or inventory of the facility. The entire nationwide security situation needs attention.

As of the end of 2005, the U.S.-Russian cooperative program to upgrade security of IRSs had focused on four activities:

1. Analysis of information available in Russian databases intended to provide inventories of the numbers, types, and locations of IRSs that are in use or in storage in Russia. These analyses are expected to lead to recommendations concerning priority sites for improved IRS protection and for consolidation of IRSs.

2. Improvement of the security and related infrastructure capabilities at Radon storage and disposal sites

3. Collection and disposal of unwanted IRSs
4. Acceleration of the decommissioning of RTGs that are or have been deployed in Russia, mainly in the Far North

DOE program officials informed the committee that their priority was to continue working in these four areas and, if resources permit, to initiate activities that will improve physical protection at health-related facilities that use high-activity IRSs.

The state enterprise Izotop is responsible for several aspects of safe handling of radionuclides in Russia, including their safe packaging and transport. Also, it is an important partner of DOE in recovering unwanted IRSs. The specific tasks assigned to Izotop under the cooperative U.S.-Russian program are as follows:

- Discover unused, poorly maintained, or abandoned radiation devices and equipment containing IRSs
- Inspect equipment and devices proposed for return and for recycling of IRSs that are not being used for their intended purposes or that have been abandoned
- Locate, dismantle, consolidate, transport, and bury IRSs in secure repositories
- Identify, plan, design, and carry out measures to modernize physical protection, control, and accounting of materials at selected sites where IRSs remain

As of December 2005, the cooperative program had recovered 1,732 IRSs with total activity of about 200,000 curies. In addition, security upgrades were installed at the Izotop handling facility. Although these achievements are welcome progress, the program thus far has only touched a very small portion of the IRSs that are unused or have become orphan sources.

More than 1,000 RTGs were produced for use in the former Soviet Union. Most of these RTGs were deployed along the coasts of Russia. Almost all were used to power remote navigational and weather stations. For example, more than 130 lighthouses in the Far North rely on RTGs for power. Most RTGs are the property of the Russian Navy, while some are under the control of the Ministry of Transportation. The RTGs typically are of very high activity and present both a safety and a security concern not only to Russia but also to its neighbors, should these devices be taken across the Russian border.

The removal of RTGs from many locations is constrained by the lack of replacement power sources. Norway has been providing solar-powered electricity generators for several years. DOE has used this experience as a base for also providing solar energy devices. Several DOE-financed pilot projects to test new solar power and wind generators are under way using navy sites. An additional pilot project will rely on commercial electrical lines for power. In some cases, Russian authorities have decided that replacement energy sources are not needed. Over the longer term, several of Russia's neighbors in addition to the United States are working with the Russian government with an eventual goal to decommission all RTGs and replace them with alternative power where needed.

The cooperative program has made good progress. The database and inventory project is beginning to provide a broad picture of the IRS situation in Russia. The rapid physical security upgrades provide much-needed and timely improvements. Some of the most dangerous IRSs contained in high-activity RTGs have been taken out of service. However, much more needs to be done by the Russian government and cooperatively to reduce the threat to both U.S. and Russian interests.

The cooperative effort has been limited in large measure by inadequate funding in both countries. Some Russian organizations that are responsible for security of IRSs have not indicated an interest in participating in the program. Of particular concern is the lack of involvement of the Ministries of Health and Social Services, Natural Resources and Energy, Agriculture, and Education and Science. All of these ministries have responsibility for stewardship of large numbers of IRSs, and the status of security procedures within the facilities of the ministries is simply not known. Also missing from active participation in the program are the hundreds of enterprises that have IRSs in their possession.

Large numbers of inadequately protected IRSs are present in many countries, and particularly IRSs for which there is no longer a need. For these unwanted IRSs, financially affordable disposal pathways often do not exist. Many IRSs are left unattended and unprotected, and they are easy prey for terrorist groups. Groups that have experience in assembling and detonating conventional bombs should be able to readily acquire the skill to handle radioactive material used in IRSs and incorporate such material in dirty bombs.

The disruption attendant to an RDD detonation could be widespread, particularly if it occurs outdoors in a densely populated urban area. The number of radiation victims might not be great. However, the likelihood of psychological impacts of a radiological attack leading to widespread fear and social disruption would be high, and the economic costs of closing off and cleaning up contaminated areas would be very significant.

The task of securing even the most dangerous IRSs in Russia is daunting. For example, hundreds of radioisotope thermoelectric generators are located in the northern reaches of the country, and the logistics to recover those that are no longer needed or could be replaced with other energy sources are formidable. Criminals have already stripped the metal off some of these RTGs, indicating the vulnerability to theft of the radioactive components as well. In addition to the problem of securing RTGs, the committee observed security deficiencies in protecting other types of IRSs of concern; and dangerous IRSs are located in hundreds of institutes, enterprises, hospitals, and other locations that are within reach of criminals. Also, the committee heard reports of unwanted IRSs being frequently discovered in abandoned facilities and in open fields.

Continued encouragement of the Russian government to address the security of IRSs more aggressively in these areas is important. Also, new opportunities for collaboration that builds on early successes have emerged.

Important problems were selected for initial program "quick fixes"—improved regional and ministry inventories of IRSs, accelerated time lines to reduce the number of vulnerable RTGs, collection and disposal of unwanted IRSs, and enhanced security at some of its storage and disposal facilities. Initial projects in each of these areas have been successfully completed. The program of quick security fixes is very important and should be continued, and the DOE leadership should expedite its implementation. Meanwhile, DOE should evaluate the effectiveness of approaches that are being used and modify them, if appropriate, to help ensure that the greatest amount of threat reduction is being achieved for the money spent. Of particular concern to the committee is the end-of-life-cycle management of IRSs that are no longer wanted, including many that have been simply abandoned. Of course, counterpart Russian organizations should be involved in evaluation efforts as well as in planning and prioritizing future activities.

A primary recommendation of the committee is that DOE develop an overall plan for the use of resources that may become available to DOE in ways that will have the maximum impact on reducing the risks attendant to inadequately secured IRSs in Russia. This plan should indicate how U.S. resources can leverage larger resources of the Russian government and thereby become an important basis for budget requests to support the program. DOE should have a comprehensive plan for all of its relevant global efforts, and within this framework, the plan for Russia should help determine the percentage of available resources that should be allocated for the Russian program.

The committee recognizes that progress toward development of a comprehensive Russian program will take time due in large measure to (1) decentralized responsibilities in Moscow and throughout the country for undertaking and financing many relevant activities, (2) chronic shortages of necessary funding either from the government or from the custodians of IRSs to correct security deficiencies, and (3) a legacy of security problems reflected in many inadequately protected IRSs, problems that are often attributable to organizations that no longer exist. Several federal laws and regulations are already in place, and specialized activities at the federal level, such as the operation of the Radon sites and the Izotop program to collect unwanted IRSs, have been established. But a comprehensive nationwide effort is still a long way off. Thus, the program should include activities to meet high-priority near-term objectives while also reflecting a vision of how best to address the security threats in the long term. Once such a program is in place, the need for DOE to continue to invest significant resources in the cooperative program should diminish. However, cooperation in this field should continue indefinitely as Russia and the United States continue to learn from each other. Although DOE's financial assistance should phase out in due time, DOE should not have an exit strategy for cooperation, because the threat of radiological terrorism will most likely persist for decades.

Sound risk analysis should be a key tool in setting priorities for the cooperative program. The committee considers the current IAEA and DOE categorizations of risks associated with IRSs to be a reasonable starting point for risk assessment. But risk depends on many factors that have not yet been adequately incorporated into national or international efforts. These factors include not only total activity and half-life but also portability, dispensability, prevalence of use, and public perceptions and fear of various radionuclides, such as plutonium. At present, only a small fraction of the millions of existing IRSs are generally considered to be high risk, but thousands of other IRSs should be of great concern when taking into account all of the risk factors. Several institutions in the United States and abroad are carrying out research on broadly based quantitative analyses of risks, and the Russian scientific community has a strong tradition in risk analysis. U.S. and Russian experts should work together to develop risk models that take into account the foregoing and other factors, which could provide an improved basis for targeting resources to problems of greatest concern.

In summary, only the Russian government has the capability to strengthen the many weaknesses in the security system for IRSs. Nevertheless, DOE and other external partners are in a good position to encourage the Russian government to develop a more comprehensive approach to ensure adequate life-cycle management of IRSs than currently exists. The development of such a comprehensive approach will be the measure of DOE's success.

NOTES

1. Committee on Opportunities for U.S.-Russian Collaboration in Combating Radiological Terrorism. 2007. U.S.-Russian Collaboration in Combating Radiological Terrorism. Washington, D.C.: National Academies Press. Available online at *www.nap.edu/catalog.php?record_id=11801*. Members of the authoring committee included John F. Ahearne (chair), director, Ethics Program, Sigma Xi, The Scientific Research Society; Laurin Dodd, managing director, Chernobyl Shelter Implementation Program (SIP), Project Management Unit, Bechtel International Systems, Inc.; Siegfried S. Hecker, director emeritus, Los Alamos National Laboratory, and visiting professor, Center for International Security and Cooperation, Stanford University; Darleane C. Hoffman, professor of the Graduate School, Department of Chemistry, University of California, Berkeley, and faculty senior scientist, Nuclear Science Division, Lawrence Berkeley National Laboratory; Roger Kasperson, research professor, George Perkins Marsh Institute, Clark University; and Leroy E. Leonard (consultant to the committee), project leader, Off-Site Source Recovery Project, Los Alamos National Laboratory.

2. Category 1 sources "if not safely managed or securely protected would be likely to cause permanent injury to a person who handled [them], or were otherwise in contact with [them], for more than a few minutes. It would probably be fatal to be close to this amount of unshielded material for a period of a few minutes to an hour." These sources are typically used in practices such as radioisotope thermoelectric generators, irradiators, and radiation teletherapy.

Category 2 sources, "if not safely managed or securely protected, could cause permanent injury to a person who handled [them], or were otherwise in contact with [them], for a short time (minutes to hours). It could possibly be fatal to be close to this amount of unshielded radioactive material for a period of hours to days." These sources are typically used in practices such as industrial gamma radiography, high dose rate brachytherapy, and medium dose rate brachytherapy.

Category 3 sources, "if not safely managed or securely protected, could cause permanent injury to a person who handled [them], or were otherwise in contact with [them], for some hours. It could possibly—although it is unlikely—be fatal to be close to this amount of unshielded radioactive material for a period of days to weeks." These sources are typically used in practices such as fixed industrial gauges involving high-activity sources (for example, level gauges, dredger gauges, conveyor gauges, and spinning pipe gauges) and well logging.

Two additional categories, 4 and 5, are also described. These contain smaller quantities of radioactive material and are generally not considered dangerous in the context of an RDD. However, when large numbers of low-activity IRSs are aggregated together and produce a total activity similar to the higher categories, a danger can exist.

17

IAEA Activities in Preventing Radiological and Nuclear Terrorism

Miroslav Gregoric, Office of Nuclear Security, Department of Nuclear Safety and Security, International Atomic Energy Agency

FOUR CONCERNS OF NUCLEAR TERRORISM

One terrorist target could be of a nuclear nature. Terrorists might acquire a nuclear weapon or nuclear material to produce an improvised nuclear explosive device (IND) or obtain other radioactive material to produce a radioactive dispersal device (RDD). If used in a city, the consequences of an IND explosion would be devastating in terms of direct human loss, while if an RDD is used in a city, the prevailing damage might not be in human casualties but rather in psychological, sociological, and economic impacts due to relocation of the population, long-term decontamination, and possible long-term health effects. Similar concern relates to the risk of a sabotage of nuclear or other facilities or transport with radioactive material.

SPECTRUM OF TARGETS

More than 120,000 nuclear weapons have probably been produced in the world in the past 60 years.[1] Many were dismantled after the cold war, but the number of existing ones is estimated at more than 25,000.[2] These weapons and related nuclear materials are clearly outside the mandate and statute of the International Atomic Energy Agency (IAEA). We must assume that all necessary measures are being carried out in states possessing these weapons and materials to address nuclear security at the level commensurate with the risk.

Worldwide inventories of plutonium and highly enriched uranium (HEU) are more than 3,000 metric tons, including both the civilian and the military sectors. There are 442 commercial nuclear power plants operating worldwide in 31 states.[3] Many of them have spent fuel storage and radioactive waste onsite. According to the IAEA database on research reactors, there are 248 research reactors in operation and 239 have been shut down awaiting further decisions.[4] HEU is still used at 60 research reactors and is present at an additional 10 research reactors that have been shut down. Within the initiatives to eliminate the use of HEU in research reactors, some operators managed to send the spent fuel and the HEU fuel back to the state of origin—the United States or Russia. Another IAEA database also shows 8 operating reprocessing plants, 18 conversion plants, 40 fuel fabrication plants, 13 enrichment plants, and 89 storage facilities.[5]

What about radioactive sources? The numbers in this area are much larger, as are the potential targets.[6] According to the IAEA Safety Standards, the sources are divided into five categories depending on the "D value" of a radioactive material known to be a dangerous source.[7] As defined in the standards, "the D value is the radionuclide specific activity of a source which, if not under control, could cause severe deterministic effects for a range of scenarios that include both external exposure from an unshielded source and internal exposure following dispersal of the source material." The number of Category 1 sources (with activity of more than 1,000 D) is estimated to exceed 10,000.[8] These include industrial and food-processing irradiation facilities, medical teletherapy units and gamma knives,[9] and radioisotope thermal generators (RTGs). The number of Category 2 sources (with activity between 10 D and 1,000 D) is estimated to be more than 100,000. These sources include industrial radiography devices and high and medium dose rate brachytherapy units. Finally, the number of Category 3 sources (with activity between 1 D and 10 D) is estimated to be more than 1,000,000; these include industrial gauges and well-logging sources. In total there are more than 3 million radioactive sources worldwide.

Operations at nuclear facilities and the use of radioactive sources also depend on the transport of nuclear or other radioactive materials. All this presents possible targets for theft of the materials, sabotage of a nuclear facility, actions aimed at inducing nuclear accidents, or other malicious acts. This situation is a subject of concern that warrants special attention.

INTERNATIONAL NUCLEAR SECURITY REGIME

Nuclear security, which involves preventing, detecting, and responding to thefts, sabotage, unauthorized access, illegal transfer, or other malicious acts involving nuclear materials, other radioactive materials, or their associated facilities, is undoubtedly the responsibility of the state. However, because of the international and transnational character of terrorism and because of possible transboundary effects of terrorist acts, an effective international system needs to be established to combat the threat of nuclear terrorism, and this system should include international cooperation and coordination. The role of the IAEA in this process is advisory; it can offer support and assistance to member states. The international nuclear security regime has slowly evolved over the past three decades. It includes the Convention on Physical Protection of Nuclear Material (CPPNM),[10] with 126 parties; the CPPNM amendment,[11] adopted by consensus in 2005 with 9 ratifications so far; the Nuclear Terrorism Convention of 2005,[12] with more than 100 signatories so far and 20 ratifications; and United Nations Security Council (UNSC) resolutions 1373 and 1540. The Code of Conduct on the safety and security of radioactive sources is an essential complement, which serves as a political commitment by states to put in place safety and security infrastructure and measures to control radioactive sources effectively. A total of 90 states have expressed such a commitment. These instruments were developed primarily to address substate actors—terrorists or criminals—and to prevent, detect, and respond to malicious acts involving nuclear and other radioactive material and facilities. The international community should strive for universal adherence to these instruments.

This framework is further enhanced by complementary safety instruments such as the Convention on Early Notification (100 parties), the Convention on Assistance in the Case of Nuclear Accident or Radiological Emergency (98 parties), the Convention on Nuclear Safety (60 parties, including all states with operating nuclear power plants), and the Joint Convention on the Safety of Spent Fuel Management and on the Safety of Radioactive Waste Management (45 parties), as well as the Code of Conduct on the Safety of Research Reactors. The latter two conventions are of particular importance because of the established periodic review mechanisms of the parties. These instruments were developed to provide a national legal infrastructure to prevent nuclear accidents and mitigate their consequences should they happen, and here again broad adherence is welcome.

SAFEGUARDS AGREEMENTS AND ADDITIONAL PROTOCOLS

An equally important enhancement to nuclear security is provided by instruments related to safeguards, such as the Nuclear Nonproliferation Treaty (NPT), safegards agreements, additional protocols (APs), and agreements on nuclear-weapons-free zones, as well as nuclear supplier rules. These instruments were developed to restrain state activities aimed at weapons development. Several of them require strong accounting and control; export and import controls; and physical protection of materials, equipment, and technology. As IAEA Director General Dr. Mohamed ElBaradei has pointed out: "The nuclear nonproliferation regime continues to face a number of challenges. I remain concerned by the fact that 30 countries have not yet fulfilled their legal obligations under the NPT to conclude and bring into force comprehensive safeguards agreements. I am also concerned by the comparatively slow progress on the conclusion and entry into force of additional protocols. To date, more than 100 countries remain without an additional protocol in force. As I have stated on many occasions, the Agency can provide no assurance with regard to countries that have no safeguards agreement, and limited assurance about the absence of undeclared nuclear material and activities with regard to countries that do not have an additional protocol in force."

The IAEA safeguards system is designed for the verification of the fulfillment of the states' commitments not to divert nuclear material from peaceful use towards nuclear weapons or other nuclear explosive devices. The agency's safeguards system includes commitments relevant to strengthening national controls over nuclear material and nuclear-related material and activities:

- **State System of Accounting for and Control of Nuclear Material (SSAC)**: Comprehensive safeguards agreements (CSAs), which are required for all nonnuclear-weapon states under the NPT and comparable nonproliferation treaties and agreements, require states to maintain effective SSACs, to ensure that nuclear material is accounted for at all times, and to record and report to the IAEA any changes in national inventories. States with APs in force are also required to provide information to the agency, inter alia, on research and development activities related to the nuclear fuel cycle but not involving nuclear material.
- **Export and import controls**: IAEA comprehensive safeguards agreements require states to report exports and imports of nuclear material to the agency. APs expand these export-reporting requirements[13] to certain specified equipment and nonnuclear material. These obligations assume that the states maintain import and export controls that enable them to report such international transfers to the agency.

CONVENTION ON THE PHYSICAL PROTECTION OF NUCLEAR MATERIAL (CPPNM)[14]

The CPPNM is included on the list of 13 legal instruments of relevance for combating terrorism adopted by the United Nations.[15] The CPPNM is the only international, legally binding undertaking in the area of physical protection of nuclear material aimed at averting potential dangers of the unlawful taking and use of nuclear material. In particular, certain CPPNM commitments are relevant to the control and protection of nuclear material:

- **Protection of nuclear material**: The CPPNM obliges contracting states to ensure the protection of nuclear material used for peaceful purposes at the levels specified in the CPPNM on their territory, ships, or aircraft during international transport. It also defines three categories of nuclear material and their corresponding levels of protection during international transport. It requires that states prohibit transport or transit unless nuclear material is protected at appropriate levels.
- **Export and import requirements**: States parties to the CPPNM commit themselves not to undertake or authorize undertaking international transports (such as the export and import of nuclear material) unless assurances are provided that nuclear material will be protected at the required levels. Parties must also apply the agreed levels of protection to nuclear material during transit from one part of their territories to another and while passing through international waters or airspace.
- **Measures to prevent, detect, and punish offenses relating to nuclear material**: Parties are obliged to make offenses relating to nuclear material punishable by appropriate penalties under their national laws and to establish their jurisdiction over such offenses. These offenses should be included as extraditable offenses in any extradition treaty existing between the parties. Parties that make extradition conditional on the existence of a treaty may consider the CPPNM as a legal basis for extradition with regard to those offenses.

CPPNM AMENDMENT OF 2005

The amendment to the CPPNM emphasizes a state's responsibility for the physical protection of nuclear material and facilities on its territory as well as for domestic and international transport of nuclear material, protection against sabotage, and securing of confidential information. The document sets physical protection objectives and fundamental principles and expands the scope of punishable acts related to nuclear material or facilities that states must prosecute. These include, among others, intentional acts involving nuclear material without lawful authority, infliction of substantial damage to the environment, smuggling of nuclear material, sabotage of a nuclear facility, organization or direction of others to commit an offense, and acts by groups of persons. In essence, the physical protection objectives are to protect against theft, locate and recover stolen material, protect against sabotage, and mitigate radiological consequences of sabotage if it occurs.[16] Fundamental principles of a state physical protection system relate to the responsibilities of a state, including responsibilities during transport, legislative and regulatory framework, establishment of a competent authority, responsibility of a license holder, security culture, a threat-based approach, a graded approach, in-depth defense, quality assurance, contingency plans, and confidentiality.

THE CODE OF CONDUCT ON SAFETY AND SECURITY OF RADIOACTIVE SOURCES

By undertaking to implement the Code of Conduct on the Safety and Security of Radioactive Sources, states are committed to reinforcing the responsibilities of manufacturers, suppliers, users, and those managing disused sources as well as those responsible for the safety and security of radioactive sources.[17] Furthermore, they are committed to establishing an effective national system of control over the management of radioactive sources; creating legislation and regulations that prescribe and assign governmental responsibilities for the safety and security of radioactive sources; and providing for the effective control of radioactive sources. In particular, under such legislation and regulations, states are obliged to include security measures to prevent, protect against, and ensure the timely detection of theft, loss, or unauthorized use or removal of radioactive sources during all stages of management. A total of 88 states have made political commitments to the Code of Conduct. Its supplementary Guidance for the Import and Export of Radioactive Sources has also been receiving growing attention, with 41 states having made a political commitment to it. In 2006, states agreed to establish a mechanism for a voluntary, periodic exchange of information among states on their implementation of the code and its guidance. This mechanism would include regional and international meetings, with an informal report summarizing the discussions. The technical meetings related to implementation of this guidance are attended by more than 50 states, including nonmember states.

UNSC RESOLUTION 1540

The UNSC Resolution 1540, which is binding for states, focuses on preventing the proliferation of weapons of mass destruction, including nuclear weapons. It specifically addresses concerns regarding terrorism and illicit trafficking, obliges all states to enforce effective measures to prevent proliferation, and specifically references the need to develop and maintain appropriate physical protection measures and accounting for nuclear material. It defines punishable acts related to proliferation that states must prosecute.

INTERNATIONAL CONVENTION FOR THE SUPPRESSION OF ACTS OF NUCLEAR TERRORISM

The nuclear terrorism convention, which was opened for signature in September 2005, is the most recent of 13 UN antiterrorism instruments. It offers a definition of acts of nuclear terrorism and a broad definition of radioactive material, thus covering radioactive dispersal devices as well as nuclear explosive devices. It details offenses relating to unlawful and intentional possession and use of radioactive material (including nuclear material), nuclear explosive devices, radioactive material dispersal, and radiation-emitting devices, as well as the use or damage of nuclear facilities. States parties are required to adopt the necessary measures to criminalize these offenses. It also includes an obligation to cooperate, share information, and inform the UN secretary general and the IAEA. States parties are "to make every effort to adopt appropriate measures to ensure the protection of radioactive material, taking into account relevant recommendations and functions of the International Atomic Energy Agency."

IAEA STANDARDS AND GUIDELINES CONTRIBUTING TO NUCLEAR SECURITY

There are several IAEA documents from the safety area contributing to nuclear security, such as the International Basic Safety Standards for Protection against Ionizing Radiation and for the Safety of Radiation Sources (Safety Series No. 115); Regulations for the Safe Transport of Radioactive Material (Safety Standard Series No. TS-R-1); Legal and Governmental Infrastructure for Nuclear, Radiation, Radioactive Waste and Transport Safety Safety Requirements (Safety Standard Series No. GS-R-1); Safety Requirements on Preparedness and Response to a Nuclear or Radiological Emergency (Safety Standards Series No. GS-R-2); Emergency Notification and Assistance Technical Operations Manual (ENATOM); and others. Also worth mentioning is the *Handbook on Nuclear Law*, covering the areas of safety, security, safeguards, and nuclear liability, which may assist states in drafting their own nuclear legislation.[18]

IAEA RESPONSE TO SECURITY THREATS

The IAEA serves as a central component of the international security infrastructure that provides the framework for cooperation. In cooperation with member states, the agency has prepared a Nuclear Security Plan 2006-2009, which was approved by the Board of Governors, and endorsed by the General Conference in September 2005, to address threats to nuclear security.[19] It is the continuation and expansion of the initial plan adopted in 2002.[20] The Office of Nuclear Security in the Department of Nuclear Safety and Security is in charge of coordinating and implementing the plan, which is divided into three areas: (1) coordination and data analysis (including the Illicit Trafficking Database), (2) prevention, and (3) detection and response. The office also coordinates activities related to safety and safeguards that contribute to security. The main such activities include promoting international instruments relating to nuclear security, establishing the international nuclear security recommendation and guidance documents, promoting the development of human resources, and providing nuclear security services, states' needs for which could be identified and addressed in the Integrated Nuclear Security Support Plans. It covers also some security upgrades for nuclear facilities and facilities with radioactive sources and is involved in combating illicit trafficking, such as through the use of border detection and monitoring equipment.

PROMOTING INTERNATIONAL INSTRUMENTS RELATED TO NUCLEAR SECURITY

The work of the IAEA has contributed to a considerable increase in the number of states that adhere to the CPPNM, which has risen from 63 states parties in 1999 to 95 in 2003 and to 126 in 2007. The need to strengthen the international physical protection regime has been widely recognized, inter alia, by the IAEA Board of Governors and the General Conference. By adopting the amendment to the CPPNM in 2005, the international community has recognized the need to strengthen the existing international legal regime in the area of nuclear security. A set of Physical Protection Objectives and Fundamental Principles has been endorsed by the IAEA Board and the General Conference and incorporated into the amended CPPNM. These objectives and fundamental principles are to be included in a security fundamentals document that will serve as further guidance to all member states. They provide a platform for national and international efforts to improve physical protection of nuclear material, particularly in their use, storage, and transport. The document underlines the need for a security culture to be implemented at all levels.

Representatives of more than 130 states have attended the agency's regional and interregional seminars on safeguards agreements, additional protocols, and the strengthened safeguards system in the past 4 years. The agency has an outreach program to encourage and facilitate states' conclusion of CSAs and APs. Several meetings have been organized internationally and regionally to promote the Code of Conduct on the Safety and Security of Radioactive Sources. Also, the IAEA has participated in several seminars aimed at promoting implementation of UNSC Resolution 1540.

GUIDELINES AND RECOMMENDATIONS: NUCLEAR SECURITY SERIES

One of the high-priority efforts of the IAEA is establishing nuclear security guidance for states to implement the conventions and other international instruments and providing supplementary measures for nuclear security at the state, regulator, and operator levels. Internationally accepted baseline documents for nuclear security are now being developed in the new Nuclear Security Series, which covers nuclear and other radioactive materials and associated facilities as well as transport. The process for the development of the Nuclear Security Series document was designed to assure high-quality consistency with other agency standards and guidance and broad international consensus through involvement of member states. The fundamentals of nuclear security will represent the top level, and recommendations will represent the second level (part of this level is currently covered by IAEA INFCIRC/225/Rev.4). The third level will consist of different implementing guidelines, which will provide guidance in areas such as design basis threats; protection against sabotage; vital area identification; protection against insider threats; identification of radioactive sources; security of radioactive sources, waste, and transport; and combating of illicit trafficking and nuclear security at major public events. Several cross-cutting areas will also be covered, including nuclear security culture, information technology security, confidentiality of information, and emergency response guidance. The first four documents in the Nuclear Security Series, covering specifications for border-monitoring equipment, detection of radioactive material in the mail, nuclear forensics, and self-assessment of nuclear facilities against sabotage, were recently published. Two more guidance documents from this series are in the final stages of publication, namely, the materials devoted to identification of radioactive sources and devices and the handbook on combating illicit trafficking of nuclear and other radioactive material.

EVALUATION AND ADVISORY SERVICES

Different nuclear security missions, evaluations, and technical visits are the IAEA's main tool for assisting states in improving their nuclear security by identifying nuclear security needs. These needs can be subsequently addressed by the state alone, with the assistance of a bilateral partner, or in conjunction with agency support, funded through the voluntary Nuclear Security Fund.

The International Nuclear Security Service (INSServ) mission serves as a flexible mechanism to help identify a state's broad nuclear security requirements for prevention, detection, and response, as well as the measures needed to meet these requirements. This is the basis for drafting the Integrated Nuclear Security Support Plan (INSSP), which, once agreed upon by the recipient state, provides an action plan to be implemented by the state, the IAEA, and an optional bilateral donor. The INSSP provides a platform for work over an extended period of time, which will cover all of the needs related to nuclear security from the prevention, detection, and response areas. It is based on the confidential INSServ report, other mission reports and technical visits, and the state's requests. It is the main tool for identifying bilateral donors and coordinating efforts with them. So far, more than 30 INSSPs have been developed, of which 12 have been transmitted formally to states for agreement.

Most relevant to the physical protection of nuclear materials and facilities are the International Physical Protection Advisory Service (IPPAS) missions. They continue to serve as the agency's main tool for evaluating existing physical protection arrangements in member states. The IPPAS missions carry out detailed reviews of the legal and regulatory basis for the physical protection of nuclear activities in the requesting state and of physical protection systems and arrangements at facilities. The IPPAS team also reviews compliance with obligations contained in the CPPNM and with guidance provided in INFCIRC/225/Rev.4 and compares their observations with international best practices. The findings of the IPPAS missions are formulated into confidential mission reports for further action. A total of 38 missions have so far been undertaken in 28 states. Specific IPPAS followup assistance, such as training, technical support, and more targeted assessments, continues to constitute an essential feature of this advisory service.

An effective SSAC is fundamental to a state's ability (1) to account for and control its nuclear material and detect possible losses or unauthorized use or removal of nuclear material and (2) to fulfill its international nuclear nonproliferation obligations. The agency offers an SSAC advisory service (ISSAS), whereby a team of experts, at the request of a state, inter alia, reviews the legal framework and regulatory, administrative, and technical systems of SSACs and evaluates the performance of those systems in meeting safeguards obligations. Recommendations are made and an action plan is formulated in cooperation with the state to improve its SSACs. The follow-up to such missions may involve, for example, assistance with equipment procurement or training of staff from the SSAC authority and from facility operators.

Two ISSAS missions and 12 SSAC evaluation missions have been undertaken at the request of member states since 2002. In addition, eight seminars and workshops relating to security and accounting of nuclear material have been held, and technical assistance (including equipment) has been provided to three states.

Furthermore, the IAEA organizes International Team of Experts (ITE) missions, composed of legal and technical experts, to advise states on adherence to and implementation of international instruments relevant to enhancing protection against nuclear terrorism. ITEs have visited 18 countries to date in Africa, Eastern Europe, Latin America, the Middle East, and Southeast Asia.

Other types of missions offer states reviews and advice regarding their regulatory infrastucture for safety and security of radioactive sources and for emergency preparedness and response arrangements.

HUMAN RESOURCES DEVELOPMENT: EDUCATION AND TRAINING

To assist states in establishing and maintaining effective nuclear security, a variety of training courses, seminars, and workshops at the international, regional, and national levels are offered. From 2002 to 2006, more than 160 training events (99 in prevention and 62 in detection and response) lasting from 3 days to 3 weeks were organized for more than 120 states and involved 3,000 participants. The target audience depends on the subject of the course or workshop but can include policy makers; nuclear regulators; facility operators; legislators; lawyers; emergency responders; police; border forces; and customs, military, and intelligence officials.

The training modules cover basic and advanced topics related to physical protection and a systematic methodology to design and evaluate physical protection systems for nuclear facilities that are effective against theft and sabotage. One module is focused on the physical protection and control of radioactive sources throughout their life cycle, while another module concentrating on transport security will be offered this year. Specialized physical protection modules include national workshops on methodology for developing the design basis threat (DBT) required to define performance targets for physical protection systems. Modules are available for protection against sabotage, vital area identification, and prevention of insider threats. Also offered are modules including hands-on training on the technical features of physical protection systems and a course to prepare national authorities for conducting inspections of physical protection arrangements. Several modules cover measures for combating and responding to illicit trafficking, border monitoring, and the use of different monitoring equipment. Some of these training courses are organized after equipment is supplied to a state.

Another example of IAEA outreach activities is the relatively new course on security of radioactive sources, which began in 2004. This course has been held in 10 states (Namibia, Algeria, Australia, South Africa, Argentina, Pakistan, Slovenia, India, Tunisia, and Syria), reaching 310 participants from 63 states. Five more training events are being implemented in 2007 (Kazakhstan, Nigeria, Spain, China, and Estonia) and will serve an addition 150 participants from 30 states. Similar statistics can be drawn for several other modules.

For university education, an academic module on physical protection, control, and accounting of nuclear material began in 2005 with IAEA support at the Sevastopol National University of Nuclear Energy and Industry (SNUNEI) in Ukraine. An essential element of the project was the collaboration of academic staff from the Moscow Engineering Physics Institute, which has had such a program since 1997, and SNUNEI. Collaboration between the Naif Arab University for Security Science (NAUSS), the King Abdul-Aziz City of Science and Technology in Saudi Arabia, and the IAEA may result in the establishment of similar educational modules on nuclear security at NAUSS in the near future.

TECHNICAL IMPROVEMENTS AND UPGRADES

IAEA is assisting states in upgrading their physical protection systems for nuclear and other radioactive material and associated facilities and in improving their border detection and monitoring equipment for customs, police, and border police that have been identified through the nuclear security services and are a part of the INSSP. As in the past, IAEA supports national efforts to increase nuclear security during major public events by providing relevant equipment and training. These technical improvements can be made with bilateral support from member states or to some extent from the IAEA Nuclear Security Fund.

CONCLUSIONS

The threat of nuclear terrorism is real. The potential targets are nuclear and other radioactive material and associated facilities or transport. As these are abundant, the possibility of terrorist acts cannot be ruled out. Threat reduction can be achieved by both eliminating materials at risk and protecting those in use.

Nuclear security is a responsibility of states themselves, but it is also subject to the emerging international nuclear security regime. States are addressing threats to nuclear and radioactive material and associated facilities and transport. Different measures are being taken to reduce the threats to and vulnerabilities of potential targets, including by means of design-basis threat reassessment and physical protection enhancements. Simultaneously, other measures have been taken to protect major infrastructure in individual states against terrorist attack, including aviation security enhancement. Information sharing, while at the same time securing confidential information at state and international levels, is slowly improving. All of this is contributing to enhanced nuclear security. Security measures are an essential element of threat reduction.

To further improve nuclear security worldwide, the international community should strive for universal adherence to international nuclear-security-related instruments and their implementation, including continued use of IAEA nuclear security advisory services and the Nuclear Security Series documents.

We should all be aware that safety, safeguards, and security are prerequisites for the sustainability and renaissance of nuclear power. IAEA has several programs and activities to cooperate with, support, and assist states in their efforts to combat nuclear terrorism and implement their international obligations, such as UNSC Resolution 1540.

NOTES

1. Norris, R., and H. Kristensen. 2006. Nuclear notebook: Global nuclear stockpiles, 1945-2006. Bulletin of the Atomic Scientists 62(4):64-67. Available online at *thebulletin.metapress.com/content/c4120650912x74k7/fulltext.pdf*. Accessed May 6, 2008.

2. The Weapons of Mass Destruction Comission. 2006. Weapons of Terror: Freeing the World of Nuclear, Biological, and Chemical Arms. Available online at *www.wmdcommission.org/files/Weapons_of_Terror.pdf*.

   International Atomic Energy Agency (IAEA). 2006. Nuclear Power and Sustainable Development. Available online at *www.iaea.org/OurWork/ST/NE/Pess/assets/06-13891_NP&SDbrochure.pdf*.

3. IAEA. Nuclear Power and Sustainable Development.
4. Research reactors database, available online at *www.iaea.org/worldatom/rrdb/*.
5. Nuclear Fuel Cycle Information System, available online at *www-nfcis.iaea.org/NFCIS/*.

6. Ferguson, C., T. Kazi, and J. Perera. 2003. Commercial Radioactive Sources: Surveying the Security Risks. Monterey, Calif.: Center for Nonproliferation Studies, Monterey Institute of International Studies. Available online at *cns.miis.edu/pubs/opapers/op11/op11.pdf.*

7. IAEA. 2005. Categorization of Radioactive Sources: Safety Guide. Safety Standards Series RS-G-1.9. Vienna: IAEA. Available online at *www-pub.iaea.org/MTCD/publications/PDF/Pub1227_web.pdf.*

8. IAEA. 1991. Nature and Magnitude of the Problem of Spent Radiation Sources. TECDOC-620. Vienna: IAEA. Available online at *www-pub.iaea.org/MTCD/publications/PDF/te_620_web.pdf.*

United Nations Scientific Committee on the Effects of Atomic Radiation. 2000. Sources and Effects of Ionizing Radiation. Available online at *www.unscear.org/unscear/en/publications/2000_1.html.*

Gonzalez, A. 2001. Security of radioactive sources: The evolving new international dimensions. IAEA Bulletin 43(4). Available online at *www.iaea.org/Publications/Magazines/Bulletin/Bull434/article8.pdf.*

9. DIRAC (Directory of Radiotherapy Centers). Available online at *www-naweb.iaea.org/nahu/dirac/default.shtm.*

10. The Convention on the Physical Protection of Nuclear Material. INFCIRC/274/Rev.1. 1980. Available online at *www.iaea.org/Publications/Documents/Infcircs/Others/inf274r1.shtml.*

11. Nuclear Security: Measures to Protect against Nuclear Terrorism. Amendment to the Convention on the Physical Protection of Nuclear Material. Report by the Director General. GOV/INF/2005/10-GC(49)/INF/6. 2005. Available online at *www.iaea.org/About/Policy/GC/GC49/Documents/gc49inf-6.pdf.*

12. International Convention for the Suppression of Acts of Nuclear Terrorism. 2005. Available online at *untreaty.un.org/English/Terrorism/English_18_15.pdf.*

13. Imports are to be reported by the state upon request of the IAEA.

14. The Convention on the Physical Protection of Nuclear Material. INFCIRC/274/Rev.1. 1980. Available online at *www.iaea.org/Publications/Documents/Infcircs/Others/inf274r1.shtml.*

15. Measures to Eliminate International Terrorism. A/RES/46/51. 1991. Available online at *www.un.org/documents/ga/res/46/a46r051.htm.*

16. First included in Nuclear Verification and Security of Material: Physical Protection Objectives and Fundamental Principles. GOV/2001/41. 2001. Available online at *www.iaea.org/About/Policy/GC/GC45/Documents/gc45inf-14.pdf.*

17. The code does not apply to nuclear material, as defined by the CPPNM, except sources incorporating plutonium-239.

18. Stoiber, C., A. Baer, N. Pelzer, and W. Tonhauser. 2003. Vienna: International Atomic Energy Agency.

19. Nuclear Security: Measures to Protect Against Nuclear Terrorism. Progress Report and Nuclear Security Plan for 2006-2009. Report by the Director General. GC(49)/17. Available online at *www.iaea.org/About/Policy/GC/GC49/Documents/gc49-17.pdf.*

20. Protection against Nuclear Terrorism: Specific Proposals. GOV/2002/10. 2002.

18

Electromagnetic Terrorism: Threat to the Security of the State Infrastructure*

Vladimir Ye. Fortov, Russian Academy of Sciences (RAS) Moscow High Temperature Institute, and Yury V. Parfyonov, RAS Institute of High Energy Densities

A real danger has arisen in recent years, namely, the possible appearance of a new variety of terrorist acts—so-called electromagnetic terrorism. This term refers to the intentional use of powerful electromagnetic pulse emitting devices or high-voltage pulse generators with the aim of disrupting the normal operations of a country's technical systems. Such systems include, for example, aircraft takeoff and landing control instrumentation; telecommunications systems; electronic devices used in managing nuclear power plant operations; systems for electricity generation, transmission, and transformation; equipment used in protecting environmentally hazardous facilities; and so forth.

The world has seen the creation of many powerful electromagnetic pulse generators capable of knocking modern electronic systems out of commission. We shall cite an example of one such piece of equipment that has been created in the laboratory. It consists of a semiconductor-based high-voltage short-pulse generator and an amplifying emitting antenna. Electromagnetic pulses with amplitude on the order of 5 kilovolts per meter and length of about 0.2 nanosecond are formed at a distance of about 10 meters from the emitter. The feature that

* Translated from the Russian by Kelly Robbins.

makes this unit unique is its compactness. We direct your attention to the maximum size of the generator, which is only about 30 centimeters. Further reductions in the size of the generator are possible, and a flat antenna may also be used.

Existing small high-voltage pulse generators make it possible to inject into data transmission chains or even into buildings' electricity supply and grounding networks pulses that are harmful to the equipment located in such buildings. They form short pulses with amplitude of 80 kilovolts, periodically repeating at a frequency of 1,000 gigahertz. Such a generator could be manufactured with a volume on the order of 500-800 cubic centimeters.

There are two possible scenarios for how acts of electromagnetic terrorism could be carried out using powerful electromagnetic sources. Option 1 would be by aiming a powerful electromagnetic field at a facility, and option 2 would be by injecting high-voltage pulses into the data transmission lines and into the electricity supply and grounding network in buildings. To assess the degree of danger presented by these scenarios, a large number of facilities were studied to determine their resistance to the impact of powerful super-broadband electromagnetic radiation and high-voltage pulse disruptions. The results of the experiments show that the intentional use of powerful pulse disruptions could lead to dangerous wide-scale consequences, such as communications breakdowns, power failures, alarm systems blockages, and so forth.

At the same time, it must be said that the designers of the most critical facilities recognize this danger and apply all possible measures to protect electronic systems from various types of electromagnetic disruptions. However, there is an enormous quantity of civilian-use electronic equipment for which there are no requirements for protection against powerful electromagnetic disruptions. Of course, if a few individual pieces of such equipment crash, there will be no serious consequences. Meanwhile, if such equipment fails on a massive scale, chaos will ensue. Therefore, systematic studies have been initiated regarding the stability of civilian-use technical systems against intentionally directed electromagnetic impacts. As an example, presented below are the results of tests on an electronic electricity-use meter and electric power line isolators.

The typical electricity meter is a complex device that includes a special integrated system, a microcontroller, power-independent memory, flow sensors, a pulse power source, an optical port, a liquid crystal indicator, a quartz generator, and a light diode. Experiments have indicated that if the meter is irradiated from a distance of 10 meters, operational failures occur. Furthermore, the personnel responsible for the electricity-use monitoring and accounting system are, as a rule, not capable of establishing the causes of the equipment failure in a timely manner or taking effective measures to eliminate them. Thus, the vulnerability of electronic electricity meters has been established experimentally. The tests have also demonstrated the fundamental possibility of intentionally disrupting their operating capacity for criminal purposes, for example, for unauthorized selection of a favorable electricity rate, and so forth. It is significant that these actions could be taken remotely and without anyone's notice.

As previously noted, electric power line isolators were among the items tested. A transformer substation would undoubtedly be a more interesting test subject; however, it is too expensive. Therefore, high-voltage isolators were selected as a focus of the experiments instead. The results of these tests are extremely interesting. It is generally believed that technical systems that include semiconductor devices are the most sensitive to the effects of pulse disruptions. As for high-voltage equipment, it is deliberately deemed resistant to such disruptions. This conclusion is based on the results of standard tests on high-voltage equipment for the impact of such disruptions in the absence of operating current. However, in actual conditions, the equipment will be simultaneously affected by both the disruptions and the operating current. Therefore, researchers concluded that special studies were needed. An experimental setup was developed for this purpose. The unit reproduces the joint action of short pulses of up to 400 kilovolts and operating electric current of up to 30 kilovolts. Electric power line isolators were tested using this setup.

Experiments on the isolators showed that with the simultaneous effects of high-voltage pulse disruptions and operating current, degradation of the isolators' electric parameters was observed along with their mechanical destruction. Such effects may lead to catastrophic phenomena in power systems similar to the widespread failure in the Mosenergo system in the summer of 2005 or the fire that broke out in the cable collector in Moscow's Central District in July 2006.

Thus, the experimental data indicate that compact super-broadband electromagnetic pulse emitters and high-voltage pulse generators could easily be used in dishonest competitive struggles, in unauthorized and unnoticed lowering of rates paid for electricity, in the organization of power system failures, and so forth. It would seem reasonable not to wait for these potential threats to be realized but instead to take timely measures to prevent them. Such measures would include evaluating the vulnerability of the most important infrastructure elements. It is also necessary to develop effective measures to protect infrastructure elements from electromagnetic terrorism. Perhaps a review and clarification will also be needed regarding rules for grounding devices and means of laying data transmissions lines, power cables, and so forth.

Because terrorism has become international in recent years and is evoking serious concern in all industrially developed countries, it would be expedient to take measures to promote international cooperation on this issue. It seems necessary to organize a joint experiment to assess the real danger of electromagnetic terrorism and develop means of protection. In addition, international and Russian standards must be developed with the aim of providing better protection for the civilian infrastructure against intentionally directed electromagnetic impacts.

19

The Phenomenon of Suicide Bombings in Israel: Lessons Learned

Detective Mordecai Z. Dzikansky, New York Police Department (NYPD) Intelligence Division, Overseas Liaison to the Israel National Police (Stationed in Israel 2003-2007), Currently Retired NYPD First Grade Detective

While terror is not new to the State of Israel, from September 2000 to 2006 a relentless terrorist campaign called the second intifada or the Al-Aqsa intifada was waged against Israel. During this period there was a surge of terrorism with new means of attack and severity previously unknown in Israel.

The focus of this report is suicide bombings within Israel, excluding the West Bank and the Gaza Strip. Although these attacks represent less than 1 percent of terrorist attacks against the State of Israel, they have resulted in 50 percent and 55 percent of terror-related fatalities and injuries, respectively. Other types of attacks include rockets, shootings, stabbings, and so forth.

Since 2000, 136 suicide bombings were carried out in Israel resulting in 516 fatalities and scores of injuries (see Figure 19-1).

As depicted in the above graph, the number of casualties resulting from suicide bombings has been on a steady decline since 2002. However, this mode of attack remains the most lethal. Additionally, despite the drop in attacks, there has been significant escalation on the ground of threat levels and the number of attempts to dispatch suicide bombers. According to Israeli security services, nearly 200 suicide attacks were thwarted over the past year and more than 300 explosive devices were detonated pre-attack.

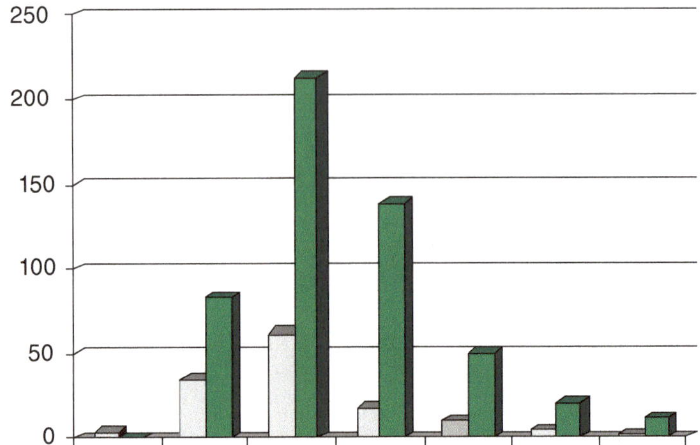

	2000	2001	2002	2003	2004	2005	2006
□ Attacks	4	35	62	18	10	5	2
■ Fatalities	0	83	213	138	50	21	11

FIGURE 19-1 Casualties from suicide bombings in Israel, 2000-2006.

WHY SUICIDE BOMBERS?

Suicide attacks are nondiscriminatory with targets, including the old, the young, the wealthy, and the poor. These attacks impact all sectors of the civilian population and have been carried out in various types of crowded public venues, including transit stations, buses, restaurants, shopping malls, nightclubs, and outdoor markets. This method of attack is adaptable, can maximize casualties, is inexpensive, and is far reaching by instilling fear in the general public. There are several factors that contribute to the effectiveness of this mode of attack:

- **Adaptability**: The bombers, who act like a human missile, can adjust their target, location, and timing of detonating their charge based on specific circumstances. Unlike a nonhuman or timed explosive, they can evaluate security, casualty potential, and positioning at their target site and modify their location and timing accordingly. There have been numerous occasions in Israel where suicide bombers either changed their target or aborted a pending attack because of circumstances at the initial site. Examples include the following:

  - **Jerusalem, September 9, 2003**: A suicide bomber attempted to enter a pizzeria (Pizza Meter) but was rebuffed by the security guard posted at the entrance, so instead the bomber entered the café (Café Hillel) next door. The guard at the entrance to the café tried to prevent the bomber from entering, but the bomber had managed to get several meters inside. Seven people were killed (including the security guard) and more than 50 were wounded.

o **Jerusalem, July 19, 2004**: The Caffit restaurant was the target of a failed attack when a would-be suicide bomber decided against carrying out his plan at the last minute because of tight security. The device that was left in the vicinity of the restaurant was recovered by the Israel National Police (INP).

o **Jerusalem, March 2002**: A waiter at the same restaurant (Caffit) noticed wires protruding from a man's backpack, cut the cable, and then tackled the would-be suicide bomber to the ground.

o **Jerusalem, August 11, 2004**: Three Palestinian bystanders were killed and six Israeli police officers were injured at the Qalandiya Checkpoint when an unattended object was detonated by remote control. The device was intended for use at a suicide bombing in Jerusalem, but the checkpoint became the modified target when the transporter deemed it impossible to pass with the device because of heavy security at the scene.

o **Netanya, July 12, 2005**: The bomber exploded his improvised explosive device (IED) approximately 30 meters away from a mall entrance, in the crosswalk. The theory is that the bomber was in the midst of a crowd in the crosswalk adjacent to the mall. Since he noticed a strong presence of security guards stationed at the mall entrance, the bomber doubled back and detonated his device in the crosswalk instead of at the mall.

Note that bombers tend to strike in areas that are familiar to them or to their handlers and will at times risk travel to that specific target (for example, attacking a target near a restaurant where the bomber was formerly employed). Terrorists may also return to the same target location if their initial attack was unsuccessful or resulted in low casualties. Examples include the following attacks:

o **Netanya, Hasharon Shopping Mall, July 12, 2005**: The first attempt resulted in two deaths. A second attempt on December 5, 2005, resulted in five deaths. In the second incident, because of security in front of the mall, the bomber detonated his explosive in the crosswalk.

o **Tel Aviv, Rosh Ha'ir Restaurant, January 19, 2006**: An explosive malfunctioned, killing only the bomber. A second attempt on April 17, 2006, resulted in 11 deaths.

o **Tel Aviv Dolphinarium Night Club, June 1, 2001**: Five months before the bombing, there was a failed attack attempted at the same location.

- **Cost versus capabilities**: Suicide bombings in a closed environment are a preferred modus operandi by terrorist groups, as only a small explosive charge is necessary to cause maximum casualties and damage. These devices are inexpensive, unsophisticated (triacetone triperoxide, or TATP), and can be prepared with easily purchased ingredients. To maximize casualties, fragmentation has been used in most of the bombings. It is the fragmentation (for example, nails, bolts, screws, and so forth) as opposed to the explosion itself that raises the number of fatalities and serious injuries. Note that devices with fragmentation are easier to detect than those without:

- o If worn, the bulkiness of the additional weight makes camouflage more difficult and more noticeable.
- o The fragmentation can be picked up by metal detectors.

- **Instills fear in the general public**: Buses during rush hour, restaurants during peak dining or social hours, markets at their busiest all these targets are selected and timed to affect the masses. During the peak period of attacks, certain changes in the behavior of segments of the population were initially noted. People avoided dining out despite an enforced law that restaurants of certain size must have security guards posted at their entrance; parents drove children to school instead of sending them on public city buses; people altered their schedules to avoid being in crowds. However, once the suicide attack became "routine" and the decline in the number of attacks began, the population appeared to return to their regular patterns. People have exhibited a strong resilience to terror attacks and seem to function better by returning to their normal routines as quickly as possible after an attack. As an example, within just 3 to 4 hours after a blast on a city bus, the bus is removed from the scene and people are seen lining up at the location waiting for the next bus.

WHO IS THE BOMBER? ANYONE

While there used to be a typical profile for the suicide bomber in Israel, the scenario changed during the second intifada. Before 2001 the suicide terrorist was typically identified as a Palestinian male, 18-28 years old, unmarried, and religious. Now there are no rules. Bombers no longer fall into a gender or age range; they include the religious and secular, professionals and laypeople, locals and foreign visitors. Examples include the following:

- **Haifa, October 4, 2003**: Twenty-one people were killed at a restaurant, including 4 children, and 60 were wounded in a suicide bombing carried out by a **female** terrorist in the Maxim restaurant in Haifa. Islamic Jihad claimed responsibility for the attack. The bomber, Hanadi Jaradat, a **29-year-old lawyer** from Jenin, blew herself up in the middle of the restaurant after completing her meal. Jaradat was the intifada's sixth female suicide bomber and the second one to do so for Islamic Jihad. Like Hamas, Islamic Jihad originally raised both religious and social objections to female bombers. The other four female bombers came from the ranks of the secular Fatah militias, Islamic organizations that have clearly overcome their religious and social objections to using women and children.

- **Hawara Checkpoint, March 24, 2004**: A **14-year-old Palestinian boy**, Hussam Abdu, wearing an explosive belt was intercepted at the roadblock, south of Nablus. Sappers (combat engineers) used a remote-controlled robot to pass scissors to the boy so that he could cut the explosive belt off his body and then safely detonated it in a controlled explosion. Abdu, from Nablus, said that he received 100 Israeli shekels (about 22 U.S. dollars at that time) to carry out a suicide attack. A Tanzim cell from the Balata refugee camp in Nablus claimed responsibility for sending the boy.
- **Jerusalem, August 19, 2003**: A Palestinian suicide bomber of Hamas' Hebron cell, apparently disguised as a Hassidic Jew, detonated himself on a no. 2 Egged bus in Jerusalem's Shmuel Hanavi neighborhood. The double-length bus was crowded with Orthodox Jewish families coming back from the Western Wall. The huge explosion caused lethal damage, killing 7 children and 16 adult civilians and wounding more than 130 people. The bomb was spiked with ball bearings designed to increase injuries on the crowded bus. Hamas claimed responsibility for the attack and identified the attacker as a **29-year-old mosque preacher** from Hebron.
- **Tel Aviv, April 30, 2003**: A suicide terrorist blew himself up at the entrance to Mike's Place, a pub-café on the Tel Aviv promenade. Three civilians were murdered and more than 50 were wounded in the attack, which was perpetrated by 22-year-old **British citizen** Asif Muhammad Hanif. A second British citizen, Omar Khan Sharif, 27, married, a resident of Derby, who was also due to have perpetrated a suicide attack, fled the scene. Khan Sharif attempted to detonate the bomb in his possession, but the bomb failed to explode. He fled the scene after discarding the bomb.

SUSPICIOUS SIGNS AND BEHAVIOR

While the actions of the suicide bomber are not humane, human bombers cannot always mask or control their naturally human traits or reactions pre-attack. It is these human behaviors that can alert security personnel and the public to pending trouble. Profiling based on nationality or race may not be acceptable; profiling based on objective criteria or suspicious behavior or both is essential to thwarting terror attacks. The human suspect may be recognized by external indicators, by unusual behavior, and by specific actions. Basic external indicators include the following:

- Inappropriate clothing for the season, place, time, or circumstance
- Holding a bag (various forms) that is incompatible with the surroundings
- Protrusions in the clothing

- Visible wires or tape
- Concealment of the hands
- Attempted adaptation to fit into the environment, but unnatural appearance—obvious or awkward attempts to blend into a crowd (dyed hair, clothes just not right)

Behavioral indicators include the following:

- Excessive nervousness
- Repeated and nervous handling of parts of clothes
- Profuse sweating
- Involuntary motions
- Apathy or gazing
- Slow-paced walking while focusing on sides, or determined walk
- Stuttering, mumbling (as if in prayer), hesitation in speaking or unresponsiveness

The following actions are also indicators:

- Attempting to stay away from security personnel
- Waiting in one specific point or observing a particular area
- Moving in search of a safe place suspiciously
- Two or more people communicating with each other while trying not to be noticed

While any of these suspicious actions warrant a search, the bottom line is that if trained security feels uncomfortable with an individual, appropriate action must be taken. It has been seen on numerous occasions that thorough checks, whether visual or by metal detectors, have been successful in identifying suicide bombers before they enter the target location. The following list includes several bombings in which casualties were lessened because of good security:

- **Kfar Saba Train Station, April 24, 2003**: A guard prevented a bomber from entering the station.
- **Mike's Place, Tel Aviv, April 30, 2003**: A guard prevented a bomber from entering the pub.
- **Jerusalem Checkpoint, May 18, 2003**: INP Border Police deterred an attack on a two-tiered bus.
- **Ha'amakim Mall, Afula, May 19, 2003**: A guard prevented a bomber from entering the mall.
- **Hillel Café, Jerusalem, September 9, 2003**: A guard prevented a bomber from entering the café.

- **Jerusalem Bus Stop, September 22, 2004**: INP at booths securing the site prevented a bomber from entering the bus.

When checks are not performed properly (such as at the Maxim Restaurant in Haifa in January 2004, when the guard did not use his metal detector), the consequences are deadly.

CAMOUFLAGE

Most IEDs used in Israeli suicide bombings have been in the form of a suicide belt or vest (with the devices attached midway to the bomber's chest) or bags (backpacks, duffels, gym bags). Recent camouflage has also included a guitar and undergarments. The bomber is typically disguised or dressed to blend in with the specific targeted environment. Following are examples from several attacks in Israel where different camouflage methods were used:

- **Bus Bombings**: In 2003 and 2004 there were eight suicide bombing attacks on public city buses. Six of the IEDs were worn in suicide belts and two were carried in bags. The bombers dressed to blend in. For example, on a bus filled with students, the bomber carried the device in a backpack; on a bus traveling through ultra-Orthodox neighborhoods, the bomber dressed in Hassidic garb.
- **Gush Katif Junction, January 18, 2005**: The IED charge was planted on the terrorist's body and held tightly against his legs from the waist down. The charge was held in place by means of black elastic sleeves that were concealed underneath his pants.
- **Stage Nightclub, Tel Aviv, February 25, 2005**: The bomber concealed an explosive-laden vest worn under an imitation leather coat. He carried the IED on his front and exploded his charge with his back to the entrance of the nightclub to maximize casualties.
- **Maxim Restaurant, Haifa, January 4, 2004**: The female bomber did not raise suspicion during her entire presence at the restaurant. She concealed an explosive-laden vest made from white cloth under her clothes. The opening was at the back. The vest had eight pockets arranged around the waist area that held the explosives. The bomber sat as one of the diners, ate her meal, then rose from the table and activated the vest.

ISRAEL'S DEFENSE: LESSONS LEARNED

Israel's vast experience in dealing with terrorism has led the country to a proactive position in developing systems to combat the terror. Their proven methods are both physical and psychological, and they involve the general public as well as trained security professionals. Israel has enacted security laws and built fences, roadblocks, and very strong intelligence capabilities. It is these methods that have led to the decline in the execution of suicide bombings.

Preventive Measures

Private security companies are responsible for securing buses, trains, and their respective stations. The INP is responsible for setting the training doctrine, overseeing its implementation, hiring the trainers, and performing background checks of private security personnel. Israel has a tremendous benefit in that most of the guards have served in the military and are willing to work at affordable rates.

Every city's central bus station and train station has mandatory security whereby **all persons** entering the facilities are physically checked via metal detector and **all bags and packages** are required to be opened and visually checked by trained security officers. High-profile stations (for example, the Jerusalem bus station) are supplemented with X-ray machines that scan every package or bag. All public schools are required to have secure physical barriers with an armed guard stationed at the entrance to the school. Trained security guards are required to be posted at shopping mall entrances, restaurants (of certain size), hotels, and most other public venues.

The presence of security guards and being screened is the norm. It is both expected and accepted that your bags and person will be screened upon entering almost any venue, from the supermarket to a movie theater.

Israel has an engaged community that is taught from a young age to be alert to suspicious objects and behavior. Security is included in the school curriculum and reminders are commonly posted in public locations.

In addition to security guards, high-profile locations are constructed with security considerations in mind. Factors include but are not limited to parking garage size, building materials, physical barriers in front of all entrances, and so forth.

The Security Fence[1]

The security fence, which is intended to prevent the infiltration of terrorists into Israel, is a key element of Israel's defense against terrorism. Until the construction of the security fence between Israel and areas of the Palestinian Authority, terrorists had almost unhindered entry into Israel because the area had no borders or natural obstacles.

According to statistics provided by the Israeli security services, since the August 2003 completion of the first section of the security fence and buffer zone, there has been a drastic reduction in the number of suicide bombings. The security fence, the buffer zone, and even sections of the fence not yet completed limit the ability of terrorist organizations to enter Israel and present operational obstacles, especially for those organizations active in northern Samaria, making it difficult for them to carry out suicide bombing attacks within Israel.

In 2006, terrorists did not cross the fence. Most of the terrorists who infiltrated to carry out suicide bombing attacks did so in areas where the fence is not complete. (During 2006, abductions and rocket fire replaced suicide bombing terrorism as the prevailing method of attack against Israel.)

As described by Israel's Ministry of Defense, the security fence is a multilayered composite obstacle comprising several elements:

- A ditch and a pyramid-shaped stack of six coils of barbed wire on the eastern side of the structure, with barbed wire only on the western side
- A path allowing Israel Defense Force (IDF) personnel to patrol on both sides of the structure
- An intrusion-detection fence in the center, with sensors to warn of any incursion
- A smoothed strip of sand running parallel to the fence, to detect footprints
- A solid barrier system: This particular design is used in a minority of cases—a total of 8 kilometers in the initial stages of the project (4 percent of the total length). Its main purpose is to prevent sniper fire into Israel and on major highways and roads. In this case a solid concrete wall resembling a highway sound barrier often used in the United States and Europe is erected. This design is used mainly along the new Trans-Israel Highway, in Bat Hefer and Matan, and in densely populated urban areas such as Jerusalem. Once the whole project is completed, the portion of the concrete sections will be 6 percent, approximately 30 kilometers.
- Various observation systems installed along the fence alerting authorities to attempted intrusions before they can be carried out
- IDF and Border Police units deployed along the security fence under the command of the IDF

Security Services

The decline in the number of suicide bombing attacks is the result of many factors, the foremost being the successful counterterrorist activities of the Israeli security forces. Three agencies play a crucial role in thwarting these attacks in Israel: the Israel Security Agency (ISA), the IDF, and the INP. The agencies realized that to be successful against the terrorists, they would need to work together to develop strong communication and share information on a real-time basis.

Israel Security Agency: The ISA's role in counterterrorism is to gather and analyze intelligence on terrorist organizations in Israel, the West Bank, and Gaza and monitor their activities. They identify terrorist cell members, location of terrorists, financing, and information on planned attacks. The ISA shares its intelligence with the IDF and INP for appropriate response.

Israel Defense Force: In addition to being responsible for monitoring the security fence, the army's role is to act on intelligence that it gathers and receives. It seeks out and destroys terrorist cells and their respective infrastructures (for example, bomb-making factories) in the West Bank and Gaza. The IDF is the front line in preventing suicide bombers from entering Israel.

Israel National Police: The INP is responsible for preventing terrorist attacks and apprehending any terrorist who has infiltrated into the country. They react to real-time intelligence and aggressively respond to handle the situation as quickly and professionally as possible. They carry out the following activities:

- Deploying police resources to search aggressively for the bomber or device and alert private security companies in the targeted region
- Setting up strategic auto checkpoints
- Securing or blocking off a suspected area, including closing down highways and blocking city entrances and exits
- Alerting the public to the situation

If a suicide bombing attack does occur, the INP is responsible for overseeing and coordinating the entire operation at the scene (emergency medical services, fire, ZAKA [voluntary emergency response teams], city services, the media, and so forth). This includes securing the area, sweeping for secondary devices, evacuating the injured, conducting forensics operations, and searching for a possible facilitator who may have dropped the bomber off at the location.

CONCLUSION

In the beginning of the second intifada, urban terrorism by means of suicide bombings was the preferred method of attack by terrorist groups. These attacks, which affected the masses and maximized casualties, were relatively inexpensive to execute, and getting to the target location was within reach. Israel's reaction to suicide bombings, the resilience of its citizens, and the country's ensuing security developments show the phenomenal capability of an urban environment succeeding against this type of terror.

The country has an engaged population—most of its private security guards have served in the IDF, and its citizens are alert and accepting of delays and checks. Security and identification of suspicious objects and behavior is taught from a very young age. Society has exhibited its ability and need to carry on after an attack.

Israel has fortified its public venues with added security measures, both structurally and with additional manpower. The addition of the security fence has made it increasingly difficult for terrorists to infiltrate into the country. Excellent intelligence and coordination among security services has greatly contributed to fighting the threat and thwarting hundreds of pending attacks. As for suicide bombings, Israel has learned to fight back. Now, on to defending itself from rocket attacks….

NOTE

1. Israel Ministry of Defense, Israel Ministry of Foreign Affairs. Israel's Security Fence 2003-2007. Information available online at *www.securityfence.mod.gov.il/Pages/ENG/default.htm.*

20

Raman Spectroscopic Detection of Chemical, Biological, and Explosive Agents[1]

Russ Zajtchuk, M.D., Professor Emeritus, Rush University Medical Center, and Gary R. Gilbert, Ph.D., Georgetown University Imaging Science and Information Systems Center, Temporarily Assigned to U.S. Army Medical Research and Materiel Command (USAMRMC) Telemedicine and Advanced Technology Research Center (TATRC)

INTRODUCTION

The real-time detection and identification of biological and chemical warfare agents as well as potential toxic industrial gases and improvised explosive devices (IEDs) is of paramount importance for protecting soldiers and first responders on the battlefield and in counterterrorism response at home.

This paper deals with the development of a chemical, biological, and explosive (CBE) detection system based on Raman spectroscopic measurements integrated to a commercially available unmanned ground vehicle (UGV) platform. Raman detection offers clear advantages over immunoassay- and DNA-based biological detection strategies, especially when configured for use on an unmanned vehicle. Raman measurements are reagentless, greatly simplifying the logistics of deployment. In addition, Raman measurements can be used to detect a broad range of CBE threats in a single measurement cycle.

By remotely guiding this sensor system to an incident area to assess soil, water, and surface contamination, exposure of personnel to a hazardous environment is prevented until the nature of the threat is fully known. Bringing the sensor to the sample also minimizes problems associated with sampling, such as cross-contamination and preanalysis decontamination, as well as the problems of disposal after the analysis is complete.

RAMAN SPECTROSCOPY FOR CBE DETECTION

Raman spectroscopy has been studied and used as a laboratory tool in chemistry for many years. It is now reaching a level of maturity that is transitioning from the laboratory to a variety of field applications. The Raman effect occurs when a photon encounters a molecule, during which time there is a chance that the energy from the scattered photon will be exchanged with vibrational bond energy of the molecule. This energy exchange manifests itself as a shift in frequency (or wavelength) in a small amount of the scattered light. Because each different chemical bond in a material causes a different frequency shift, the pattern of these shifts, known as the Raman spectrum, is unique to that material.

The Raman spectrum reveals the molecular composition of materials, including the specific functional groups present in organic and inorganic molecules. The Raman spectrum is a characteristic property of a material, just like its color or melting point, and can be used to determine the presence or absence of the material.

The detector will always be measuring an agent spectrum in the presence of the spectrum from the background or from any other material that may be present. Fortunately, in most real-world situations, the ratio of the amount of agent to the background and any other materials has significant spatial variation. This variation in composition leads to slightly different Raman spectra from different areas on the sample surface. These differences in spectra provide enough information for chemometric processing of the data, allowing identification of the agent and the background materials. The Raman Bio Identification (RBI) system computer receives a command initiated by the operator to acquire and analyze a sample from the UGV central processing unit (CPU). Using software previously developed by the ChemImage Corporation of Pittsburgh, Pennsylvania, up to 19 spectra are acquired from the sample. The laser power is typically 12 milliwatts, resulting in a laser power density of 86 watts per square centimeter. The exposure time used to acquire the spectra in testing was 10 seconds, and each measurement is the product of 10 averages (see Figures 20-1 and 20-2).

RAMAN BIO IDENTIFICATION (RBI) DETECTOR

The overall concept of the RBI robot demonstration system (Wolverine) was to integrate an RBI point sensor (the RBI head) onto a UGV manipulator arm, and then couple it to an instrument package mounted on the main chassis of the UGV. The coupling of the point sensor is accomplished through both electrical and fiber optic cables running along the manipulator structure.

The RBI detector is a Raman point sensor or a Raman proximity detector. To operate, it needs to be close but not necessarily touching the surface to be measured. The RBI detector contains subsystems to allow targeting of the head (video camera and fine-positioning system), laser illumination of the sample to induce

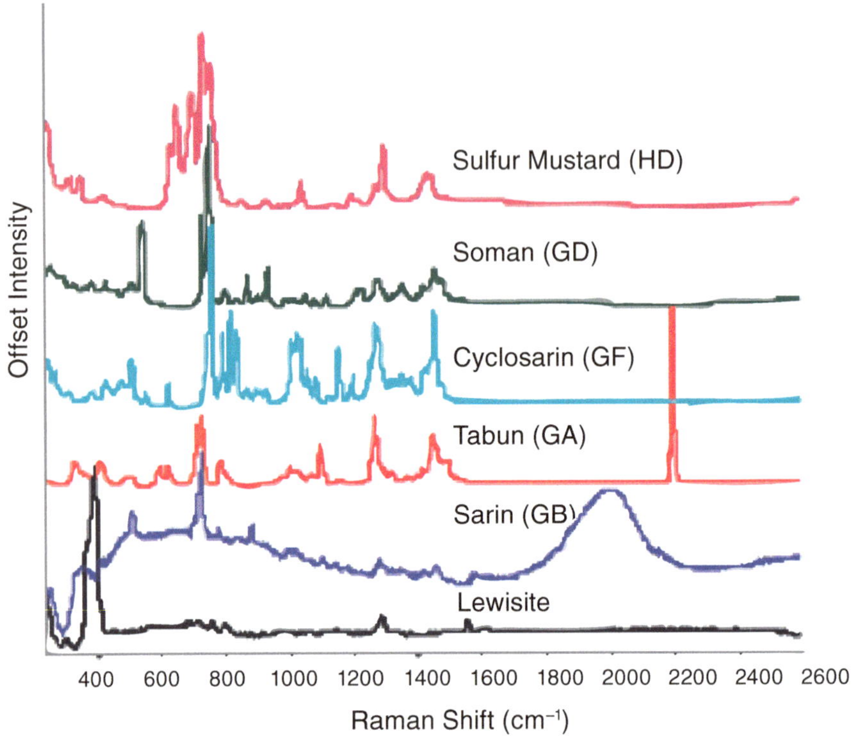

FIGURE 20-1 Raman spectra of several chemical warfare agents.
SOURCE: Gardner et al., 2007.

the Raman effect, optics to collect and focus scattered light, a fiber optic bundle to transport the scattered light to a spectral analyzer (spectrometer subsystem), and a system computer to provide control and communication (see Figure 20-3).

Using a single laser illumination spot and a single spectrometer, the RBI detector can produce up to 19 spatially resolved spectra from a sample region of interest. These spatially resolved spectra can be processed using a mixture analysis algorithm coupled with library searching to provide robust identification of threat and nonthreat materials present in complex environmental samples.

UGV INTEGRATION

The Wolverine is controlled using a radio frequency link between the robot and the operator control unit (OCU). A payload interface allows control and data transmission to and from the RBI system through this wireless interface. The

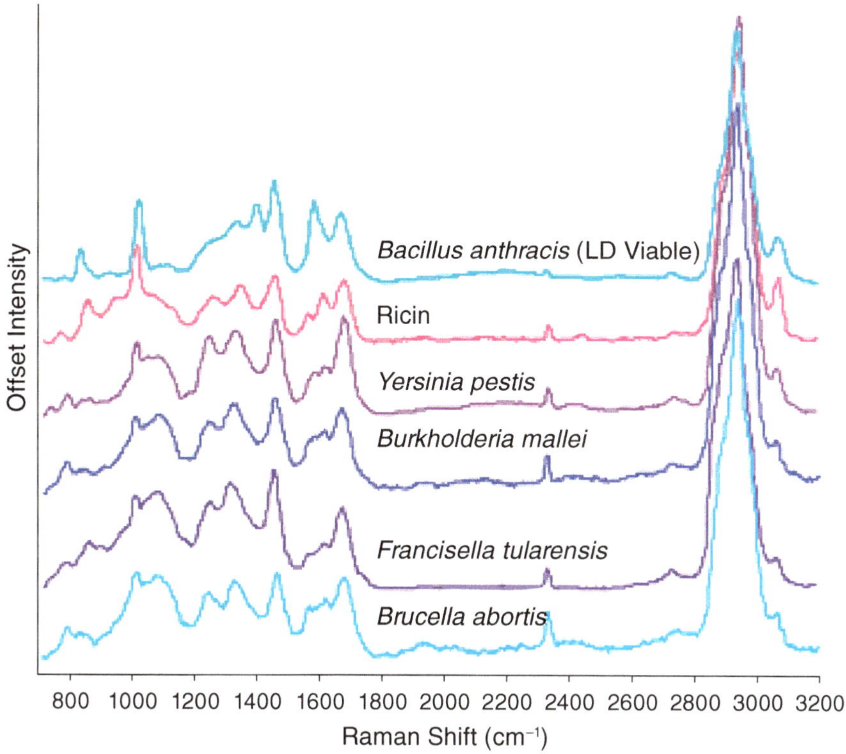

FIGURE 20-2 Raman spectra of selected biothreat agents.
SOURCE: Gardner et al., 2007.

operator has access to this video stream and can use it for fine control of the RBI head.

The proof of concept consisted of placing a biological toxin simulant, ovalbumin, on a flattened sheet of galvanized iron air-duct material to provide a constant background for the measurement. The operator then moved the UGV close to the sample area and used the manipulator to position the RBI detector head directly over the sample. The fine-adjustment system in the RBI head was used to set the collection lens of the detector at the proper distance from the sample through commands from the OCU.

Once the detector was positioned, the operator started the analysis. The analysis consisted of a 3-minute wait period, during which the native fluorescence of the ovalbumin was quenched by the laser excitation. Next, a 1-minute acquisition of the 19 spatially resolved Raman spectra was taken. This set of spectra was preprocessed and analyzed.

To confirm proper spectral performance, the spectra were averaged, prepro-

FIGURE 20-3 RBI detector mounted on the UGV.
SOURCE: Gardner et al., 2007.

cessed, and compared against the library spectrum for ovalbumin. There was good agreement between the RBI and library spectra, which confirmed the accuracy of the RBI detector.

Once the analysis and reporting of results was complete, the operator had the option of taking another measurement or moving the detector to another sampling location.

LASER-INDUCED BREAKDOWN SPECTROSCOPY (LIBS) COMBINED WITH RAMAN SPECTROSCOPY

Laser-induced breakdown spectroscopy (LIBS) is a detection method that can be used to identify chemical and biological hazards in bulk and on surfaces. It is relatively straightforward, requires no sample preparation or consumables, is sensitive, uses only a small sample substrate, is fast (subsecond), is field portable, and can be miniaturized.[2] A laser is aimed at a target and is used to quickly heat the target into a plasma plume. The resulting atomic emissions from the plasma are read by a broadband spectrometer. By measuring the relative intensities of emission spectra peaks and their specific pattern, LIBS can monitor all chemical elements present in a sample, at the same time, with a single laser shot. LIBS has been used to identify substances both in their solid and liquid forms and on top of soil samples.

There is significant improvement in detection of chemical, biological, and explosive materials when fusing data from two orthogonal technologies, LIBS and Raman, resulting in a dramatic decrease in false positive rates.

The joining of LIBS and Raman into a single sensor unit makes great sense, as both techniques can use the same laser system and same spectrometer. Also, the advanced chemometrics for spectral data analyses can be shared. LIBS and Raman are true orthogonal technologies because LIBS keeps track of the elemental composition of the sample, or target (that is, establishes its stoichiometry) exceptionally well, while Raman provides unique molecular signature information, both of which facilitate the determination of whether the target material is hazardous. Moreover, LIBS and Raman are "universal" sensors that can be applied to a very wide range of materials analyses, both hazardous and benign. The ultimate launching of a LIBS–Raman sensor payload on a robotics platform will lead to unprecedented capabilities for field applications in both proximity and standoff sensing modes.

PHOTON SYSTEMS DEEP UV RAMAN AND FLUORESCENCE DETECTOR PROJECT

Photon Systems, Inc., and the NASA Jet Propulsion Laboratory are collaborating on a project to develop an advanced, miniature, low-power, reagentless, robot-mounted, laser-based instrument for real-time detection and classification of trace concentrations of biological and chemical agents on surfaces. A combined sensor employing deep ultraviolet (UV) laser-induced native fluorescence (UVLINF) with deep UV resonance Raman spectroscopy (UVRRS) was selected for this project. This instrument is a deep UV laser consuming less than 5 watts of battery power that simultaneously generates Raman scattering and excites native fluorophores contained within microorganisms and many organic and inorganic materials. Using an onboard real-time algorithm, the UVLINF and UVRRS data are processed to identify and classify contaminants in less than 1 second. Simultaneous multiband fluorescence and Raman sensor outputs are processed using neural net algorithms to classify contaminant organic and inorganic materials.

During Phase I of this effort, Photon Systems successfully designed, fabricated, and demonstrated a capability of solar-blind, standoff detection and classification of trace amounts of biological and chemical contaminants and explosives on surfaces at working distances of 1-3 meters, significantly greater than the proposed instrument standoff goal of 5-30 centimeters. This optical instrument is not affected by ambient lighting, because of a combination of its operation in the deep ultraviolet range and the use of pulse-gated detection for background reduction. This is a very important feature for use of the instrument under natural or artificial lighting. The entire instrument was integrated into a single, robot-arm-mounted package with a weight of 5 pounds and power consumption of 5 watts, rather than the original two-part instrument with one-half on the robot arm and one-half in the robot body.

The instrument includes onboard microprocessors and firmware for controlling the laser as well as each detector and for performing a variety of computational and self-calibration tasks. The overall data processing using chemometric software for detecting and classifying unknown surface contaminants was performed by remote computer via a wireless link.

CONCLUSION

A mathematical model was constructed to describe the performance of the detector system. This mode was evaluated using a simulant for anthrax, *B. thuringiensis* (*Bt*) spores, and a biological toxin simulant, ovalbumin. This modeling confirmed the feasibility of the design for biothreat agent detection.

While Raman identification of agents requires a spectra library for pattern-matching recognition of molecular structure, it can still identify new, unknown agents (such as recombinant chemical agents or explosive structures or genetically altered organisms) by flagging for further investigation the spectra not already present in the library, especially if they closely resemble the spectra of a known agent or class of known agents.

Warning the user of the presence of an unknown substance that could possibly be a threat is of great value to the war fighter or emergency responder. Such warnings could then be incorporated into standard operating procedures for donning mission-oriented protective posture gear or other personal protective equipment. Ultimately, the suspect spectra would be added to the library as an unknown spectra associated with a potential hazard.

Work is continuing to refine the RBI detector hardware and software to allow integration on a wider class of UGV platforms and to optimize system operation for field use.

ACKNOWLEDGMENTS

This work was supported by the U.S. Army Medical Research and Materiel Command under Contract No. W81XWH-06-C-0010.

DISCLAIMER

The views, opinions, and findings contained in this paper are those of the authors and should not be construed as an official Department of the Army position.

NOTES

1. Gardner, C. W., et al. 2007. Demonstration of a robot-based Raman spectroscopic detector for the identification of CBE threat agents. Manuscript submitted for the Twenty-Fifth Army Science Conference, sponsored by the Assistant Secretary of the Army for Acquisition, Logistics and Technology, November 27-30, 2007, Orlando, Florida. Defense Technical Information Center Report AD-A481010, available online at *hdl.handle.net/100.2/ADA481010.*
2. Batavia, P., and R. Watts. 2004. Collaborative robots design considerations. Presentation at the National Defense Industrial Association Fourth Annual Intelligent Vehicles Systems Symposium, Traverse City, Michigan, June 2004.

21

The U.S. Department of Homeland Security Science and Technology Directorate

John O'Neil, U.S. Department of Homeland Security

On September 11, 2001, terrorists hijacked four civilian airliners in the United States and turned them into weapons. Government reaction at all levels to the specific activities of these three separate terrorist acts was swift. The government then undertook a more deliberate effort to protect the homeland security of the country. President Bush appointed Tom Ridge, former governor of Pennsylvania, to direct the effort.

In January 2003, 22 agencies[1] were combined to form the 15th U.S. Cabinet department, the Department of Homeland Security (DHS). This development was a monumental undertaking in bureaucratic organization. All 22 agencies had developed their own unique cultures, and it was no easy task to combine them into one department.

In August 2006, Rear Admiral Jay Cohen (U.S. Navy, Retired) became the under secretary of homeland security for science and technology. Capitalizing on his 6 years of experience as the chief of naval research, one of his first acts was to reorganize the Science and Technology Directorate (see Figure 21-1) to meet the demands of the department to develop protective technologies and to meet the following goals:

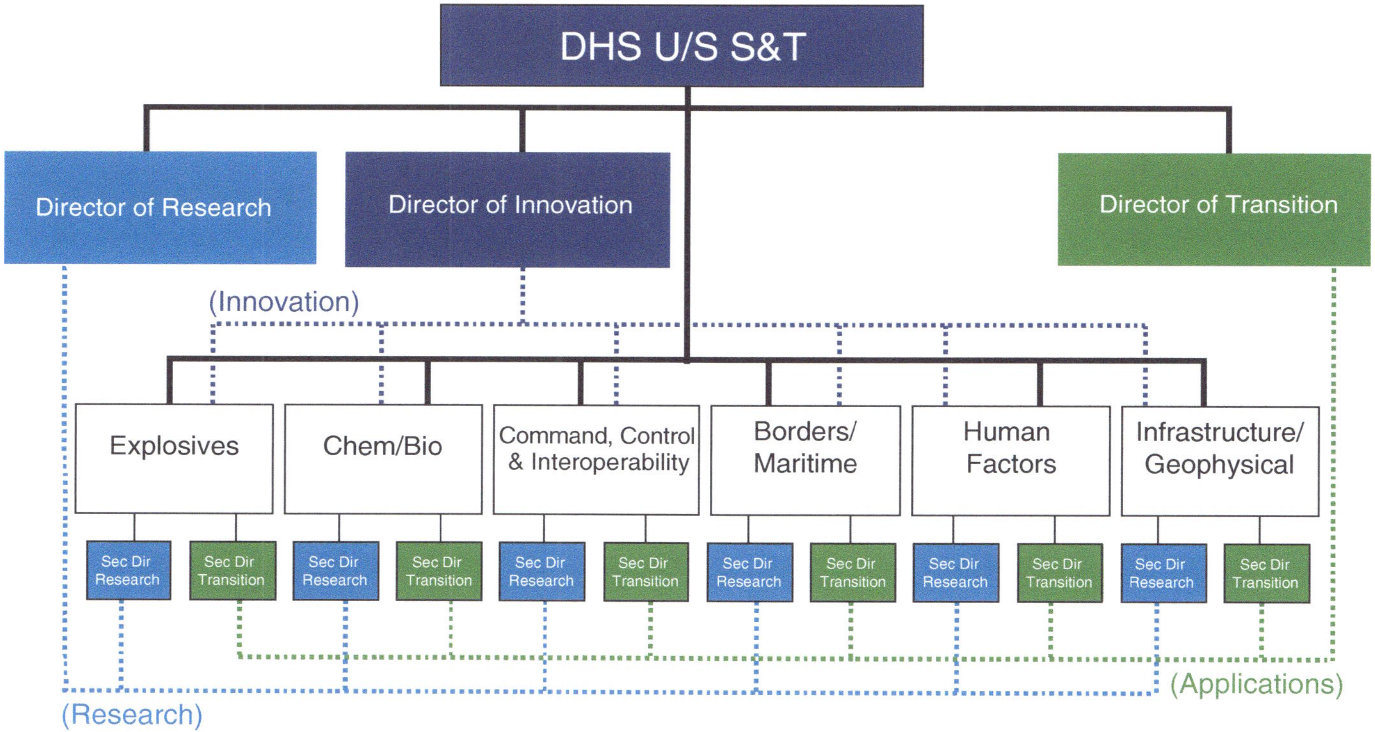

FIGURE 21-1 Science and Technology Directorate at the Department of Homeland Security.

- Accelerate delivery of enhanced technological capabilities to meet requirements and fill capability gaps to support DHS agencies in accomplishing their mission
- Establish a lean and agile world-class science and technology management team to deliver the technological advantage necessary to ensure DHS mission success and prevent technology surprises
- Provide leadership, research and educational opportunities, and resources to develop the necessary intellectual basis to enable a national science and technology workforce to secure the homeland

These three goals guided the realignment process to provide the nation with a robust capability in science and technology for homeland security applications. These goals facilitate integrated, innovative solutions to homeland security challenges.

The Science and Technology Directorate made significant strides in the first nine months of Under Secretary Cohen's tenure. A major accomplishment was to put in place the following:

- A framework for a customer-focused, output-oriented science and technology management organization
- A senior leadership team and key organizational components
- Six research divisions and a director for each
- Three portfolio directors: research, innovation, and transition
- Directors of test, evaluation, and standards and special programs
- Science and Technology Directorate liaison offices embedded in Europe, the Americas, and the Asia-Pacific area
- A communications department
- 340 employees relocated into new working groups

The core organization of the newly aligned Science and Technology Directorate includes six technical divisions that are linked to the three research and development investment portfolio directors in a matrix management structure. The technical divisions are as follows:

- Explosives
- Borders and Maritime Security
- Chemical and Biological
- Human Factors
- Command, Control, and Interoperability
- Infrastructure and Geophysical

The three portfolio directors coordinate aspects of the investment strategy with the technical divisions. The portfolio directors are

- director of research,
- director of transition, and
- director of the Innovation/Homeland Security Advanced Research Projects Agency (HSARPA).

The two top priorities of the Science and Technology Directorate are interoperability and countering improvised explosive devices (IEDs), with a goal of predicting, detecting, destroying, and defeating IEDs at a minimum distance of 100 meters.[2]

The Science and Technology Directorate develops and manages an integrated program of science and technology, from basic research through technology transition to customers. The customers are the operating components of DHS; state, local, and tribal governments; first responders; and private sector entities. Scientists and engineers in the many disciplines relevant to homeland security manage the program.

The investment portfolio of the Science and Technology Directorate is balanced around risk, cost, impact, and time to delivery to produce capabilities of high technical quality that are responsive to homeland security requirements. As shown in Figure 21-2, it consists primarily of product transition, innovative capabilities, and basic research.

The basic research portfolio addresses the long-term research and development needs of the DHS mission. Discovery and invention lead to future capa-

Balance of Risk, Cost, Impact, and Time to Delivery

Product Transition (0-3 yrs) • Focused on delivering near-term products/enhancements to acquisition • Customer IPT controlled • Cost, schedule, capability metrics	**Innovative Capabilities (1-5 yrs)** • High-risk/High payoff • "Game changer/Leap ahead" • Prototype, Test and Deploy • HSARPA
Basic Research (>8 yrs) • Enables future paradigm changes • University fundamental research • Gov't lab discovery and invention	**Mandated Spending (0-8+ yrs)** • Required by Administration (HSPDs) • Congressional direction/law

Customer Focused, Output Oriented

FIGURE 21-2 DHS Science and Technology Directorate investment portfolio.

bilities and mobilize the capabilities, talents, and resources of the Homeland Security Centers of Excellence, Department of Energy national laboratories, and DHS laboratories to address the long-term research and development needs of DHS in sciences of enduring relevance. This type of focused, protracted research investment has the potential to lead to paradigm shifts in the nation's homeland security capabilities.

The HSARPA component looks for the "high-risk but high-payoff" investments to produce "game-changing or leap-ahead" solutions. It seeks such solutions through two programs:

1. The Homeland Innovative Prototypical Solutions (HIPS) are designed to deliver prototype-level demonstrations of game-changing technologies in 2 to 5 years. Projects are moderate to high risk, with high payoff.
2. The High Impact Technology Solutions (HITS) are designed to provide proof-of-concept answers within 1 to 3 years that could result in high-payoff technology breakthroughs. While these projects entail considerable risk of failure, they offer the potential for significant gains in capability.

The transition component seeks to identify potential technology solutions that can be delivered within 3 years. It employs a concept called the Integrated Product Team (IPT) to bring together the customer, acquisition partner, science and technology leaders, and the end users to identify customers' needs by identifying operational capability gaps and requirements. The IPT makes informed decisions about technology investments for near-term capabilities to address the requirements.

The Science and Technology Directorate uses the Small Business Innovative Research Program (SBIR) to look for solutions outside government. Through SBIR, it challenges small businesses to bring innovative homeland security solutions to reality from the private sector.

In summary, the Science and Technology Directorate's mission is to protect the homeland by providing federal, state, local, and tribal officials with state-of-the-art technology and resources. Under Secretary Cohen has changed the organization to accomplish this mission. His goal is for the directorate to become a full-service organization that is customer focused and output oriented, cost effective, efficient, responsive, agile, and flexible.

NOTES

1. The 22 agencies that became part of the Department of Homeland Security in 2003 were the U.S. Customs Service; the Immigration and Naturalization Service; the Federal Protective Service; the Transportation Security Administration; the Federal Law Enforcement Training Center; part of the Animal and Plant Health Inspection Service; the Office for Domestic Preparedness; the Federal Emergency Management Agency; the Strategic National Stockpile and National Disaster Medical System; the Nuclear Incident Response Team; the Domestic Emergency Support Teams; the National Domestic Preparedness Office; the Chemical, Biological, Radiological, and Nuclear Countermeasures Programs; the Environmental Measurements Laboratory; the National Biological Weapons Defense Analysis Center; the Plum Island Animal Disease Center; the Federal Computer Incident Response Center; the National Communications System; the National Infrastructure Protection Center; the Energy Security and Assistance Program; the U.S. Coast Guard; and the U.S. Secret Service. See the DHS Web site "History: Who Became Part of the Department?" at *www.dhs.gov/xabout/history/editorial_0133.shtm*. Accessed May 23, 2008.

2. For a comprehensive list of the 12 priority Science and Technology Directorate functional areas identified by DHS, see DHS Science and Technology Directorate. 2008. High-Priority Technology Needs. Available online at *www.dhs.gov/xlibrary/assets/High_Priority_Technology_Needs.pdf*. Accessed July 17, 2008.

Appendixes

Appendix A

Agenda

PLENARY SESSION OF WORKSHOP
MARCH 21, 2007

Opening Remarks

Konstantin Frolov, Institute of Machine Sciences (Russian cochair)
Siegfried Hecker, Stanford University (American cochair)

Reports of Working Groups and of Cooperative Activities

Bioterrorism Working Group

- David R. Franz, Midwest Research Institute
- Sergei Netesov, Novosibirsk State University

Transportation Vulnerabilities Working Group

- George Bugliarello, Polytechnic University
- Konstantin Frolov, Institute of Machine Sciences

Energy System Vulnerabilities Working Group

- Siegfried Hecker, Stanford University

Cooperation in Addressing Radiological Terrorism

- Siegfried Hecker, Stanford University
- Leonid Bolshov, Nuclear Safety Institute

Presentations

International Atomic Energy Agency (IAEA) Activities in Preventing Radiological and Nuclear Terrorism, Miroslav Gregoric, Office of Nuclear Security, Department of Nuclear Safety and Security, IAEA

Struggle with the Threat of New Infectious Diseases in the Twenty-First Century, Sergei Netesov, Novosibirsk State University

Strategy of Russia in Countering Terrorism in Current Times, Valentin Sobolev, National Security Council of Russia

Basic Tendencies in the Development of Global Terrorism, Raphael Perl, Organization for Security and Cooperation in Europe

Unified Russian Governmental System in Countering Terrorism and the Tasks of the National Anti-Terrorism Committee, O. M. Zhidkov, National Anti-Terrorism Committee

Scientific-Technical Activities of the U.S. Department of Homeland Security in the Struggle with Terrorism, John O'Neil, U.S. Department of Homeland Security

Methods of Combating Suicide Bombers, Mordecai Dzikansky, Detective and Representative to the Israel National Police, New York City Police Department (retired)

Closing Session

Discussion of Presentations

Plans for Future Meetings

Closing Remarks

- Siegfried Hecker, Stanford University
- Konstantin Frolov, Institute of Machine Sciences

WORKING GROUP ON BIOTERRORISM

Presentations and List of Additional Participants

Emerging Viral Infections on the Territory of the Asian Part of Russia, Sergei Netesov, Novosibirsk State University

Modulation of Innate Immunity to Protect Against the Biological Weapons Threat, Vitaly Zverev, Andzharpidzhe Institute for Antiviral Preparations

Species Neutral Disease Surveillance and Other Opportunities in International Biosecurity, David R. Franz, Midwest Research Institute

Raman Bio-Identification Robot Analysis and Reporting of Results to Operator Control Unit, Russ Zajtchuk, Chicago Hospitals International

International Cooperation on Disease Surveillance in the Context of the Biological Weapons Convention, Michael Moodie, Consultant

Commentaries

- Vadim Ivanov, Director, Russian Academy of Sciences (RAS) Institute of Bioorganic Chemistry
- Valery Galchenko, Director, RAS Institute of Microbiology
- Aleksandr Ginsburg, Vice President of Russian Academy of Medical Sciences (RAMS)
- Viktor Zavorokhin, Deputy Director, Office of Federal Medical and Biological Agency
- Mikhail V. Ugrumov, Councilor of the Presidium of the RAS on Foreign Affairs, Professor and Chief of both the Laboratory of Hormonal Regulations, RAS Institute of Developmental Biology, and the Laboratory of Neurohistology, RAMS Institute of Normal Physiology

Site Visits

Federal Medical-Biological Agency (Moscow)

- Mikhail Kiselev, Deputy Director
- Valery Dobritsa, Director of Research Institute of Highly Pure Biopreparations
- Vladimir Romanov, Chief State Sanitary Doctor for Organizations and Territories
- Gennady Galkin, Office for Organization of Scientific Research

- Natalya Kalinina, Independent Consultant

Center for Hygiene and Epidemiology of Moscow

- Aleksandr Ivanenko, Chief Doctor, Management and Administration
- Aleksandr Mizgailov, Deputy Chief Doctor, Management and Administration
- Irina Litkina, Department of Epidemiological Supervision
- Natalya Volkova, Manager of Epidemiological Department
- Nina Salova, Manager of Department, Microbiology Laboratory
- Vitaly Pugachov, Department of Activity Planning and Organization

Research Institute for Influenza (St. Petersburg)

- Mariana Yerofeeva, Chief of Laboratory for Trials of New Antiviral Preparations
- Lyudmila Tsybalova, Deputy Director for Science

Research Institute of Highly Pure Biopreparations (St. Petersburg)

- Sergei Ketlinsky, Deputy Director for Science
- Aleksandr Ishchenko, Head of Protein Biochemistry Laboratory
- Andrei Simbirtsev, Head of Laboratory of Immunopharmacology

Public Health Center of the City of St. Petersburg

WORKING GROUP ON TRANSPORTATION VULNERABILITIES

Presentations

Scientific Basis for Countering Urban Transportation Terrorism, Konstantin Frolov, Institute of Machine Sciences

Control and Supervision as a Prerequisite for Ensuring Safety of Transportation Systems, Vladimir Chertok, Federal Transportation Supervision Service

Counterterrorism Awareness Training for Mass Transit: Employees, Customers, Police Officers, Joseph Bober, New Jersey Transit Police Department

National and International Priorities in Countering Transport Terrorism, Vladimir Lopatin, Research Institute of Intellectual Property

Strategic Approach in Protecting Transportation Facilities, Mordecai Dzikansky, New York Police Department (retired)

Application of Magnetic Inductive Tomography for Control of Passenger Flow, Vladimir Cherepenin, Institute for Radio Engineering and Electronics

Measures and Technologies for Ensuring Blast Proofing and Blast Resistance of Transportation, Industrial, Energy, and Civil Facilities, Adolf Mishuev, Blast Resistance Research and Development Center

Transportation Planning that Takes into Account Evacuation Concerns, John Falcocchio, Polytechnic University

Countering Terrorism in Organizing the Operation of Complex Transportation Systems, Viktor Dosenko, International Academy of Transport

General Issues in Preventing and Responding to Transportation Incidents, George Bugliarello, Polytechnic University

Special Features and Damaging Factors of Technological Terrorism, Nikolai Makhutov, Institute of Machine Sciences

Site Visits

Research Institute for Civil Defense and Disaster Management (EMERCOM)

- S. Kachanov, Deputy Director for Civil Defense and Disaster Management
- Sergei Todoseichuk, Chief of Emergency and Rescue
- Valery Akimov, Chief of Center for Strategic Research
- A. P. Popov, Chief of Information Technologies Department
- I. V. Sosunov, Specialist

Research Institute for Fire Protection (EMERCOM)

- Irek Khasanov, Chief of Research Institute for Fire Protection
- Aleksandr Matyushin, Deputy Chief Research Institute for Fire Protection

Northwest Regional Center for Civil Defense, Emergency Situations, and Elimination of Consequences of Natural Disasters, and Center's Field Station for Search and Rescue (EMERCOM, St. Petersburg)

Municipal Security Committee (St. Petersburg)

Emergency Response Center (St. Petersburg)

WORKING GROUP ON ENERGY SYSTEM VULNERABILITIES

Presentations

Development of Strategic Master Plan for Submarine Decommissioning as Example of an Approach to Decisions on Global Safety Issues, Ashot Sarkisov, Nuclear Safety Institute

Security of Pipelines, Sergei Serebryakov, RAS Institute of Oil and Gas Problems

Underwater Technologies for Liquified Natural Gas and Strengthening of Global Energy Safety, Vyacheslav Kuznetsov, Russian Research Center—Kurchatov Institute

Electromagnetic Terrorism: Threat for the Energy Infrastructure of a State, Yury Parfyonov, Scientific Association for High Temperatures

Industry-Sponsored Studies of Vulnerabilities of U.S. Power Systems, Siegfried Hecker, Stanford University

Security of Nuclear Power Plants, John Ahearne, Sigma Xi (presented by Siegfried Hecker, Stanford University)

Current Russian Requirements on Protection of Nuclear Power Facilities, Boris Krupchatnikov, Rostekhnadzor

Homeland Security and Energy Facilities, Drew Lieb, New Jersey State Police

U.S. Department of Homeland Security's Interest in Science and Technology to Counter Terrorism, John O'Neil, U.S. Department of Homeland Security

Safety of Gas Pipelines, Vitaly Gridin, Oil and Gas Research Institute

Strategic Approach to Protecting Energy Facilities, Rafael Perl, Organization for Security and Cooperation in Europe

Site Visits

Central Production and Control Department, Gazprom

Rosenergoatom Crisis Center

Institute of Information and Automation of the Russian Academy of Sciences (St. Petersburg)

Center of Environmental Security of the Russian Academy of Sciences (St. Petersburg)

Appendix B

Recent Russian and International Publications of Interest*

Dvorkin, V., ed. 2002. Terrorism in a Metropolis: Assessment of Risks and Defenses. National and Global Security Series. Moscow: PIR Center. Available online at *www.pircenter.org/data/publications/nz21.pdf*.

Frolov, K. V., et al. 1998. Functioning and Development of Complex Economic, Technological, Energy, Transport, Communications, and Public Utility Systems. From the Series Security of Russia: Legal, Socioeconomic, and Scientific-Technical Aspects. Moscow: Znanie.

Interaction among Administrative Agencies, Institutions, and Specialized Units in Eliminating the Consequences of Terrorist Acts Involving the Use of Pathogenic Biological Agents and Hazardous Chemical Substances: Methodological Recommendations MR 0100/3556-04-34. Official Publication: State Sanitary-Epidemiological Regulations of the Russian Federation. 2005. Moscow: Sanepidmedia (in Russian).

International Atomic Energy Agency (IAEA). 2004. Regulations for the Safe Transport of Radioactive Material, Safety Standards Series No. TS-R-1. Vienna: IAEA. Available online at *hazmat.dot.gov/regs/intl/st1_rev.pdf*. Accessed May 9, 2008 (in English).

*Copies of these publications are available for review at the National Academies, Office for Central Europe and Eurasia.

International Terrorism in the Commonwealth of Independent States: Materials from a Roundtable (Moscow, November 2002). 2003. Moscow: Antiterrorism Center of the Commonwealth of Independent States.

Kiselev, O. I, and V. I. Pokrovsky, eds. 2006. The Russian Federation National Program for Influenza Pandemic Preparedness. Moscow and St. Petersburg: Ministry of Health and Social Development and Russian Academy of Medical Sciences (in Russian and English).

Lopatin, V. N., ed. 2006. Terrorism and Security on Transport. Proceedings of the Fifth International Scientific and Practical Conference, Moscow, February 8-9, 2006. Moscow: PROEKSPO (in Russian with introduction in English).

Luzhkov, Yu. M., et al. 1998. The Security and Sustainable Development of Major Cities. From the Series Security of Russia: Legal, Socioeconomic, and Scientific-Technical Aspects. Moscow: Znanie (in Russian).

Moscow City's Emergency Prevention and Response System. 2003. Moscow: Main Administration of the Moscow City Department of Emergency Situations (in English and Russian).

On the Sanitary-Epidemiological Welfare of the Population of Moscow in 2002. 2003. Moscow: Center for Sanitary-Epidemiological Protection in Moscow, Ministry of Health (in Russian).

Russian Academy of Sciences Nuclear Safety Institute. 2005. Opportunities for U.S.-Russian Cooperation in Combating Radiological Terrorism. Unpublished report prepared under contract between the institute and the National Academies (in English).

Russian Academy of Sciences Scientific Research Center for Ecological Safety (SRCES). 2006. Trouble Shooting of Underground Pipelines of House Heating Systems by Satellite and Airborne Infrared Thermal Survey. St. Petersburg: SRCES (in English).

Shoigu, S. K., ed. 2005. Emergency Services of Russia, 1990-2005. Moscow: Ministry for Civil Defense, Emergency Situations, and Liquidation of Consequences of Natural Disasters (in Russian).

Technological Terrorism: Materials from the Scientific-Practical Conference Problems of Technological Terrorism and Methods of Preventing Terrorist Threats, November 27-28, 2003. 2004. Moscow: Russian Academy of Sciences and the Ministry of Extreme Situations (in Russian).

Vishnevsky, Yu. G., et al. 2003. Regulation of Nuclear and Radiation Safety. From the Series Security of Russia: Legal, Socioeconomic, and Scientific-Technical Aspects. Moscow: Znanie, Scientific-Technical Center for Nuclear and Radiation Safety, and Gosatomnadzor (in Russian).

Vorobyov, Yu. L., et al. 2006. Risk Analysis and Security Problems. Part One: Fundamentals of Security Analysis and Regulation. From the Series Security of Russia: Legal, Socioeconomic, and Scientific-Technical Aspects. Moscow: Znanie.

Appendix C

Russia's Counterterrorism Strategy

Valentin A. Sobolev, Deputy Secretary,
Security Council of the Russian Federation

MAIN THREATS TO INTERNATIONAL SECURITY

- Regional crises and conflicts
- Terrorism and various forms of political and religious extremism
- Separatism
- Illegal drug trade
- Environmental and technogenic disasters
- Threats of the spread of weapons of mass destruction
- Organized crime

CHARACTERISTICS OF TERRORISM IN 2005

- **Increased rates of growth in the number of terrorist acts** (according to U.S. data, more than 10,000 terrorist acts have been committed worldwide in the past three decades)
- **Rise in the level of organization** (during the twentieth century, terrorism developed from a lone terrorist model to transnational terrorist organizations such as al Qaeda)
- **Improved material-technical and financial support** (from the dagger and pistol to colossal bombings and the possible use of weapons of mass

destruction, from meager financial resources to funding streams totaling in the millions)

- **Increased scope of terrorist activity** (from single locations of crimes to the seizure of entire cities, countries, and regions)
- **Increased severity of consequences and number of casualties**
- **Expanded social base of terrorism**
- **Increased number of trained fighters** equipped at the highest technical level

CHARACTERISTICS OF MODERN TERRORISM IN 2007

- **Expanded geography and internationalization**: More than 50 countries have experienced the consequences of terrorist acts, including Iraq, India, Indonesia, Colombia, Pakistan, Afghanistan, Russia, Israel, Great Britain, and Egypt.
- **Increased danger to society**: The number of acts carried out by suicide attackers (shahids) has increased fivefold in the past 3 years. In 2006 alone, more than 15,000 terrorist acts were carried out worldwide, killing or injuring more than 90,000 people.
- **Expanded social base and involvement of significant masses of the population in extremist activities**: This leads to the creation of a broader infrastructure for terrorist organizations and brings the ethnonational factor to bear, which in turn creates a significant degree of uncertainty about potential sources of the terrorist threat and forms and means of operation by terrorists.
- **Rise in linkages between terrorism and the ethnoreligious factor**: Primarily this refers to the aim of certain branches of Islam to create individuals who are psychologically prepared to commit violent acts "in the name of Allah" to achieve political goals, such as overthrowing unfavorable secular regimes and establishing a government according to Islamic doctrines.
- **Increased level of organization and unification of terrorist organizations both within individual countries and on an international level**: Terrorists are creating a system of control with unified leading entities that plan their actions. Terrorist groups that are similar in their ideological, political, nationalistic, religious, and separatist positions are holding councils and meetings, bringing together the leaders of the largest groups.
- **Formation of three threatening hotbeds for the spread of terrorism in the world**: These regions where armed conflicts are prevalent include "Palestine-Israel," Iraq, and Afghanistan.
- **Trend toward the expansion by jihadists of their supply, funding, and personnel bases beyond the bounds of the Muslim world**: Latin America is gradually becoming a promising source from which the Islamic fighters may augment their ranks. This trend will continue, according to the predictions of Western experts in 2007. In addition to the Middle East and Western Europe, we

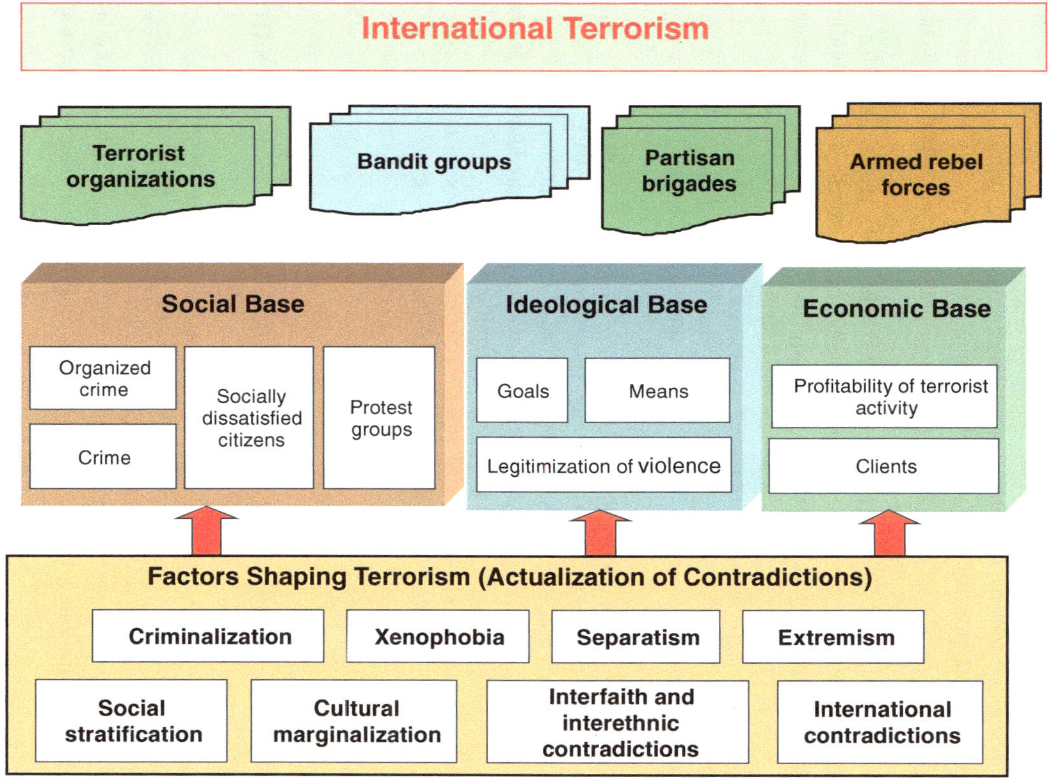

FIGURE C-1 International terrorism.

should expect increased activity by jihadists in Bosnia, Kosovo, India, Bangladesh, Indonesia, Australia, the Philippines, the Fergana Valley, and the Xinjiang Uighur Autonomous Region of the People's Republic of China.

- **Continuing material and financial support for terrorism**: The main source of financing for terrorism today comes from control of the drug trade, racketeering, prostitution, arms trade, contraband, gambling, and so forth.
- **Use of modern technologies by terrorists**: Terrorists are striving to gain access to weapons of mass destruction and their components. We must organize efforts to counter nuclear terrorism, cyberterrorism, ecoterrorism, agroterrorism, and radiological terrorism.
- **Increased ties between terrorism and the drug trade**: Terrorism is increasingly active in the so-called instability belt, which extends from the Philippines and Indonesia through the Indian subcontinent, Central Asia, the Caucasus, and the Middle East up to the Serbian border of Kosovo. The flow of drugs from Afghanistan has become global in nature. It may be stated that the efforts of the international community and the Afghan authorities to counter the production and illegal trade of narcotics are still not having the necessary effect.

Success in the struggle against terrorism is unimaginable without a clear and universally accepted international strategy. Governmental and social structures, official networks, and the media must join forces. The foundation for such an endeavor was laid by Resolution 1373 and other decisions by the United Nations Security Council, but additional efforts are currently needed.

The main elements of Russia's strategy for international cooperation to counter international terrorism and extremism include the following:

- The United Nations
- The Group of Eight (G8)
- Expanded contacts and cooperation on antiterrorism with the North Atlantic Treaty Organization (NATO), within the framework of the Russia-NATO Council
- Enhanced regional antiterrorist cooperation with nearby countries, primarily through the Commonwealth of Independent States, the Collective Security Treaty Organization, and the Shanghai Cooperation Organization
- Establishment of cooperation on countering new challenges and threats with the Association of Southeast Asian Nations and the Asian Regional Forum

Figure C-1 illustrates the new transnational ideology and practice of asymmetric violent resolution of contradictions on a global level.